Transforming Agricultural Education for a Changing World

Committee on a Leadership Summit to Effect Change in Teaching and Learning

Board on Agriculture and Natural Resources
Board on Life Sciences

Division on Earth and Life Studies

NATIONAL RESEARCH COUNCIL
OF THE NATIONAL ACADEMIES

THE NATIONAL ACADEMIES PRESS
Washington, D.C.
www.nap.edu

THE NATIONAL ACADEMIES PRESS 500 Fifth Street, N.W. Washington, DC 20001

NOTICE: The project that is the subject of this report was approved by the Governing Board of the National Research Council, whose members are drawn from the councils of the National Academy of Sciences, the National Academy of Engineering, and the Institute of Medicine. The members of the committee responsible for the report were chosen for their special competences and with regard for appropriate balance.

This study was supported by the U.S. Department of Agriculture, the W.K. Kellogg Foundation, the National Science Foundation, the Farm Foundation, and the American Farm Bureau Foundation for Agriculture. Any opinions, findings, conclusions, or recommendations expressed in this publication are those of the author(s) and do not necessarily reflect the views of the U.S. Department of Agriculture, National Science Foundation, or any of the other organizations that provided support for the project.

This material is based upon work supported by the Cooperative State Research, Education, and Extension Service, U.S. Department of Agriculture, under Award No. 200638837-03642; the National Science Foundation under Grant No. DUE-0540637; and the W.K. Kellogg Foundation under Award No. P0116528.

International Standard Book Number-13 978-0-309-13221-3 (Book)
International Standard Book Number-10 0-309-13221-5 (Book)
International Standard Book Number-13 978-0-309-13222-0 (PDF)
International Standard Book Number-10 0-309-13222-3 (PDF)

Additional copies of this report are available from the National Academies Press, 500 Fifth Street, NW, Lockbox 285, Washington, DC 20055; (800) 624-6242 or (202) 334-3313 (in the Washington metropolitan area); Internet, http://www.nap.edu.

Printed in the United States of America

THE NATIONAL ACADEMIES

Advisers to the Nation on Science, Engineering, and Medicine

The National Academy of Sciences is a private, nonprofit, self-perpetuating society of distinguished scholars engaged in scientific and engineering research, dedicated to the furtherance of science and technology and to their use for the general welfare. Upon the authority of the charter granted to it by the Congress in 1863, the Academy has a mandate that requires it to advise the federal government on scientific and technical matters. Dr. Ralph J. Cicerone is president of the National Academy of Sciences.

The National Academy of Engineering was established in 1964, under the charter of the National Academy of Sciences, as a parallel organization of outstanding engineers. It is autonomous in its administration and in the selection of its members, sharing with the National Academy of Sciences the responsibility for advising the federal government. The National Academy of Engineering also sponsors engineering programs aimed at meeting national needs, encourages education and research, and recognizes the superior achievements of engineers. Dr. Charles M. Vest is president of the National Academy of Engineering.

The Institute of Medicine was established in 1970 by the National Academy of Sciences to secure the services of eminent members of appropriate professions in the examination of policy matters pertaining to the health of the public. The Institute acts under the responsibility given to the National Academy of Sciences by its congressional charter to be an adviser to the federal government and, upon its own initiative, to identify issues of medical care, research, and education. Dr. Harvey V. Fineberg is president of the Institute of Medicine.

The National Research Council was organized by the National Academy of Sciences in 1916 to associate the broad community of science and technology with the Academy's purposes of furthering knowledge and advising the federal government. Functioning in accordance with general policies determined by the Academy, the Council has become the principal operating agency of both the National Academy of Sciences and the National Academy of Engineering in providing services to the government, the public, and the scientific and engineering communities. The Council is administered jointly by both Academies and the Institute of Medicine. Dr. Ralph J. Cicerone and Dr. Charles M. Vest are chair and vice chair, respectively, of the National Research Council.

www.national-academies.org

COMMITTEE ON A LEADERSHIP SUMMIT TO EFFECT CHANGE IN TEACHING AND LEARNING

JAMES L. OBLINGER (*Chair*), Chancellor, North Carolina State University, Raleigh
JOHN M. BONNER, Executive Vice President, Council for Agricultural Science and Technology, Ames, Iowa
PETER J. BRUNS, Vice President for Grants and Special Programs, Howard Hughes Medical Institute, Chevy Chase, Maryland
VERNON B. CARDWELL, Morse-Alumni Distinguished Teaching Professor, University of Minnesota, St. Paul
KAREN GAYTON COMEAU, Past President, Haskell Indian Nations University, Lawrence, Kansas (Retired)
KYLE JANE COULTER, Past Deputy Administrator, U.S. Department of Agriculture, Washington, D.C. (Retired)
SUSAN J. CROCKETT, Vice President and Senior Technology Officer, Health and Nutrition, General Mills, Inc., Minneapolis, Minnesota
THEODORE M. CROSBIE, Vice President of Global Plant Breeding, Monsanto Company, Ankeny, Iowa
LEVON T. ESTERS, Assistant Professor of Agricultural Education and Studies, Iowa State University, Ames
A. CHARLES FISCHER, Past President and Chief Executive Officer, Dow AgroSciences LLC, Indianapolis, Indiana (Retired)
JANET A. GUYDEN, Associate Vice President of Research and Dean of Graduate Studies, Grambling State University, Grambling, Louisiana
MICHAEL W. HAMM, C.S. Mott Professor of Sustainable Agriculture, Michigan State University, East Lansing
MICHAEL V. MARTIN, Chancellor, Louisiana State University, Baton Rouge
SUSAN SINGER, Laurence McKinley Gould Professor of the Natural Sciences, Carleton College, Northfield, Minnesota
LARRY VANDERHOEF, Chancellor, University of California, Davis
PATRICIA VERDUIN, Vice President, Global Research and Development, Colgate-Palmolive Company, Piscataway, New Jersey

STAFF

ADAM P. FAGEN, Study Director
ROBIN A. SCHOEN, Director, Board on Agriculture and Natural Resources
KAREN L. IMHOF, Administrative Assistant
PEGGY TSAI, Program Officer
NORMAN GROSSBLATT, Senior Editor

BOARD ON AGRICULTURE AND NATURAL RESOURCES

BOARD ON LIFE SCIENCES

AMANDA P. CLINE, Senior Program Assistant
REBECCA L. WALTER, Senior Program Assistant

Preface

In April 1991, leaders in the higher education community, business, industry, and public agencies met at the National Academy of Sciences in Washington, DC, for a national conference on the changes needed to meet the challenges of undergraduate professional education in agriculture. The meeting, "Investing in the Future: Professional Education for the Undergraduate," was organized by the National Research Council's Board on Agriculture[1] with support from the U.S. Department of Agriculture (USDA) and its office of Higher Education Programs in the Cooperative State Research Service, and emanated from a series of discussions of USDA Project Interact: An Integrated Curriculum Development Action Plan. The proceedings of the meeting were published by the National Academies as *Agriculture and the Undergraduate* in 1992. Although the report did not offer recommendations, it did contain a large number of ideas that were presented at the conference and has served as a source of inspiration.

Since 1991, however, a lot has changed. Universities are different, careers are different and constantly evolving, and even the meaning of the term *agriculture* has changed. Moreover, what students expect, what is expected of them, and the need for a scientifically educated population have expanded.

Over the last several years, the Academic Programs Section of the National Association of State Universities and Land-Grant Colleges (NASULGC)—now known as the Association of Public and Land-grant Uni-

[1]The Board on Agriculture and Natural Resources, which has overseen the present study, is the successor to the Board on Agriculture.

versities (APLU)[2]—has discussed what is needed to update the agriculture curriculum and prepare agriculture students for a 21st-century workplace. NASULGC approached the National Academies with the idea for a seminal event (a "leadership summit") and a National Research Council report that would draw wide attention to undergraduate education in agriculture. Conversations with various stakeholders revealed that many had similar concerns, and several in the federal government (USDA and the National Science Foundation) and the private sector (the W.K. Kellogg Foundation, the Farm Foundation, and the American Farm Bureau Foundation for Agriculture) signed on to support the project.

A committee was appointed by the National Academies in early 2006 to consider what an undergraduate education in agriculture should comprise to prepare a flexible and well-prepared workforce. The committee includes academic leaders in the land-grant university system, senior managers in food and agriculture industries, experts in science education and faculty development, faculty with experience in these topics, and representatives of professional societies.

The committee was charged with investigating how institutions of higher education can improve the learning experience for students at the intersection of agriculture, environmental and life sciences, and related disciplines. It looked at innovations in teaching, learning, and the curriculum that could be used to prepare a workforce that would meet the needs of employers and the entire community.

Central to the committee's work was organizing the Leadership Summit to Effect Change in Teaching and Learning. That meeting, held October 3–5, 2006, at the National Academy of Sciences, brought together over 300 representatives of academe, business and industry, government, professional societies, and other stakeholders. Participants ranged from university presidents to undergraduate students and from agribusiness CEOs to entry-level employees, including many in between. Sessions focused on agriculture and on education.

Most academic participants in the summit came as part of small institutional teams. The committee recommended that academic institutions applying to participate develop a team of four individuals which included a senior administrator whose responsibility extended beyond the college of agriculture (such as a provost or a dean of undergraduate education) and a

---

[2]NASULGC changed its name to APLU, effective April 1, 2009. Throughout the report, the organization will be referred to as NASULGC when referring to events and actions prior to that date.

person with responsibility for undergraduate education in agriculture (such as an associate dean for academic programs); additional team members included faculty, students, and other administrators.

Industry teams were encouraged to include senior managers in both research and development and human resources, department managers, recruiters, and others. Professional societies were represented by both staff and members, including executive directors and chairs of education-related committees.

In designing the agenda for the summit and preparing this report, the committee has had input from many people. Several representatives of the committee met with the NASULGC Academic Programs Section in February 2006 to hear their thoughts on the most important issues of concern. The committee held a planning meeting in May 2006 at which it heard from representatives of several project sponsors (USDA, the Farm Foundation, and the American Farm Bureau Foundation for Agriculture), the U.S. Environmental Protection Agency, NASULGC, and people associated with the 1991 meeting and the associated 1992 proceedings.

The committee met again before and immediately after the leadership summit to begin identifying the major themes for inclusion in its report. After the summit, several small committee working groups developed sections of the report. Those sections and the overall conclusions and recommendations were discussed at the committee's final meeting in April 2007 and in later teleconferences and other discussions.

Acknowledgments

The committee is grateful to the many hundreds of persons who have shared their insight, experiences, and suggestions on the issues being addressed in this project. The committee thanks the several hundred stakeholders who participated in the Leadership Summit to Effect Change in Teaching and Learning on October 3–5, 2006, especially the speakers, breakout leaders, and panelists: Thomas M. Akins, C. Eugene Allen, Caitilyn Allen, John C. Allen, Jerry Bolton, Gale A. Buchanan, Ralph J. Cicerone, M. Suzanne Donovan, Jay Ellenberger, Frank Fear, W.R. (Reg) Gomes, The Honorable Mike Johanns, Nicholas R. Jordan, Wynetta Y. Lee, Jose P. Mestre, Jay Moskowitz, Jeanne Narum, Marion Nestle, Paul Roberts, Gary Rodkin, Sally L. Shaver, Vanessa Sitler, Paul D. Tate, Andrew L. Waterhouse, Robin Wright, and Robert T. Yuan. We also thank the students from the University of Maryland, College Park, who served as note-takers.

The Academic Programs Section of the National Association of State Universities and Land-Grant Colleges—now the Association of Public and Land-grant Universitites (APLU)—initiated recent discussion of the issues in this report and deserves much credit for pushing these topics to the front of the national agenda. Ian Maw, vice president, Food, Agriculture, and Natural Resources at APLU, deserves particular recognition for helping to make this study come about.

The committee has benefited from the insight of those who have been thinking about the issues for many years. C. Eugene Allen (University of Minnesota), Karl G. Brandt (Purdue University), and Paul Williams (University of Wisconsin–Madison) provided their perspectives on the 1991 meeting and the last 15 years; their insight helped to guide the committee in organizing a summit—and writing a report—that would build on the earlier discussions. Representatives of several of the project sponsors also helped in shaping the content of the meeting and in suggesting topics for inclusion in the report. Additional thanks go to others who provided information to the committee to assist in drafting the report, including many of the speakers at the Summit as well as Patti Clayton, Diane Ebert-May, Christine Pfund, and Janelle Tauer. Finally, in addition to the original authors of the background papers (Appendix C and D), Joe Hunnings (University of Vermont) provided assistance in updating the data in Appendix C for final publication and has been added to the author list.

This report has been reviewed in draft form by individuals chosen for their diverse perspectives and technical expertise in accordance with procedures approved by the National Research Council Report Review Committee. The purpose of this independent review is to provide candid and critical comments that will assist the institution in making its published report as sound as possible and to ensure that the report meets institutional standards for objectivity, evidence, and responsiveness to the study charge. The review comments and draft manuscript remain confidential to protect the integrity of the deliberative process. We wish to thank the following individuals for their review of this report:

George Acquaah, Bowie State University
C. Eugene Allen, University of Minnesota
Carol Balvanz, Iowa Soybean Association
Ellen Bergfeld, American Society of Agronomy, Crop Science Society of America, and Soil Science Society of America
Alan R. Berkowitz, Cary Institute of Ecosystem Studies
Donald L. Cawthon, Tarleton State University
W.R. (Reg) Gomes, University of California (retired)
Larry Gundrum, Kraft Foods (retired)
Robert Haselkorn, University of Chicago
Molly Jahn, University of Wisconsin–Madison
Donald L. Johnson, Grain Processing Corporation (retired)
Neil Knobloch, Purdue University
Karen S. Kubena, Texas A&M University
Gordon E. Uno, University of Oklahoma

Although the reviewers listed above have provided many constructive comments and suggestions, they were not asked to endorse the conclusions or recommendations, nor did they see the final draft of the report before its release. The review of this report was overseen by **Melvin D. George**, University of Missouri (retired), and **Frederick A. Murphy**, University of Texas

Medical Branch at Galveston. Appointed by the National Research Council, they were responsible for making certain that an independent examination of this report was carried out in accordance with institutional procedures and that all review comments were carefully considered. Responsibility for the final content of this report rests entirely with the authoring committee and the institution.

Contents

Tables, Figures, and Boxes

TABLES

FIGURES

BOXES

Summary

During the next ten years, colleges of agriculture will be challenged to transform their role in higher education and their relationship to the evolving global food and agricultural enterprise. If successful, agriculture colleges will emerge as an important venue for scholars and stakeholders to address some of the most complex and urgent problems facing society. Such a transformation could reestablish and sustain the historical position of the college of agriculture as a cornerstone institution in academe, but for that to occur, a rapid and concerted effort by our higher education system is needed to shape their academic focus around the reality of issues that define the world's systems of food and agriculture and to refashion the way in which they foster knowledge of those complex systems in their students. Although there is no single approach to transforming agricultural education, a commitment to change is imperative.

WHAT IS THE URGENCY?

Our world is changing at an increasing pace and unleashing a complicated set of problems and opportunities. For example, it has always been acknowledged that the growing world population exerts a looming pressure on the global food supply, but few anticipated how population growth would converge with rising incomes in the developing world to create an unprecedented demand for more food, especially animal protein. It is now far from clear if an expansion of animal and grain production, and its associated impact on the environment and land use, both in the United States and in other agricultural countries, are even capable of satisfying the need for nutritious food in the long term. This is made even more difficult, because another new demand—that for biofuels—has placed further pressure on supplies. We are only beginning to understand the meaning of the emerging bio-economy for world food and energy security, and how this development in our agricultural system can be achieved more sustainably, if at all. It is not an exaggeration to observe that world stability depends on reliable supplies and stable prices for food and energy, which are now linked in agriculture, and on the preservation of the natural resource base that underpins all economic activity and the global way of life in the long term. Is the next generation of leaders in agriculture prepared to address these critical demands on our agricultural systems? Can we sustain the educational institutions that will prepare the leaders of tomorrow?

The search for solutions to meet urgent food, fiber, and fuel needs is complicated by issues that are beyond the control of a single nation or even one economic sector. A decade ago, the reality of climate change and the prospects for serious, negative impacts on food production and on human and animal health were not recognized. Now, the expansion of world food production must occur in potentially difficult environmental conditions at the same time that the agricultural enterprise is increasingly obligated to mitigate its own greenhouse gas emissions. Addressing the relationship of climate change and agriculture will require the sharing of insights by a diverse set of experts and actors, from scientists and engineers to regulators and policymakers both in- and outside of agriculture. Where will we find individuals with the knowledge and ability to communicate across disciplinary domains on these issues, and who will bring them together to explore solutions?

The collective global enterprise that supports and carries out the production of plants and animals and that buys, processes, and distributes agricultural products to the world's markets is huge and growing. In concert with the public institutions that both support and regulate their activities, hundreds of thousands of local, national, foreign, and multinational firms—some of them large and integrated operations, others small and specialized—orchestrate a level of economic activity that is staggering in its magnitude, breadth, and diversity of scale. It would take pages to list all the niches that have emerged in the agricultural enterprise beyond the farmer—this workforce includes scientists, seed suppliers, crop insurers and bankers, food chemists, ethanol producers, packaging engineers, food safety and quality control experts, agro-ecologists, veterinarians, meat inspectors, risk assessors, contract negotiators, shippers, grocery and retail store suppliers, institutional food buyers, and on and on. This collection of individuals, businesses, and institutions must work together across disciplines, language gaps, physical distances, and national differences to achieve their goals.

Often they must grapple with issues beyond their immediate control—such as the spread of avian flu, a plant disease outbreak, or the introduction of melamine—that threaten food supplies and shake the confidence of their buyers and consumers. Because agriculture is affected by so many conditions, its participants must always be prepared to react, to adapt, and to think ahead. How do we recruit and cultivate the workforce of the future for this diverse and dynamic universe of enterprises?

As the largest food producer in the world, the U.S. agricultural system has benefited from years of investment in technological improvements to agriculture, entrepreneurial and well-developed markets for agricultural inputs and products, public support of agricultural businesses, and a natural environment that is conducive to growing plants and animals. But because agricultural production is embedded in social and natural systems, it is affected by changing circumstances in those systems, such as increasing international competition in agricultural products, changing consumer demands and expectations of agriculture and food, declining levels of public research support, evolving immigration and labor policy, growing demands to regulate the environmental externalities of agriculture, and emerging constraints of the natural resource base. In addition, rising rates of obesity are leading to increased incidence of preventable disease while structural and economic issues affect access to fresh fruits and vegetables in many communities. How will we respond to these challenges? Do we have a pool of individuals capable of navigating us through these changing waters?

If colleges of agriculture believe they provide the logical focus for preparing these individuals, then a greater effort is needed to be successful in taking on this responsibility. Herein is the challenge to colleges and departments of agriculture: to establish a place at the forefront of academe where students and scholars are prepared to learn about the complexities of agriculture and grapple with its evolution and change, and in so doing, find their opportunity to contribute as leaders and participants in the agricultural enterprise. Only this will ensure a system of agriculture and of agricultural education that is sustainable, able to adapt to and thrive in constantly changing times.

WHY UNDERGRADUATE EDUCATION IN AGRICULTURE MUST CHANGE

It is not simple to keep up with the evolving nature of the agricultural enterprise. It requires a much more dynamic approach to the curriculum and teaching than most colleges of agriculture have developed. Moreover, many of the colleges have not fully recognized that changes have also taken place in their own educational institutions. The pool of potential candidates for the agricultural disciplines is no longer a relatively homogenous group of young people who grew up on farms. That number is diminishing, while the student population has grown increasingly diverse in terms of age, background, and culture. The diverse and broader student body is generally unaware of the multi-dimensional and challenging nature of the agricultural disciplines and the exciting career opportunities open to them, despite evidence that many students have an interest in a variety of scientific, business, economic, environmental, and social issues related to food and agriculture. The problem is that educators have not helped students to make the connection between those issues and a degree in agriculture.

In many ways, agriculture is intertwined with other disciplines in the natural and social sciences, with agriculture professionals using similar approaches and systems as those in other fields. Agriculture now so thoroughly combines basic and applied aspects of the traditional STEM disciplines of science, technology, engineering, and mathematics that the acronym might rightly expand to become STEAM, joining agriculture with the other fundamental disciplines.

Many faculty members do not have experience in the broader food and agricultural enterprise (let alone in traditional production) that would enable them to give students a "real-world" interpretation of the ideas, concepts, and skill sets they need to acquire to be effective in the diverse agricultural workplace. And few academic institutions support faculty and students in gaining real-world experience as part of learning; neither are there sufficient resources for faculty to experiment with how to refashion the way they teach or provide experiences that reflect the challenges that food and agriculture graduates will need in their future careers.

This report describes aspects of the undergraduate educational experience in food and agriculture that need to be created, strengthened, or modified. If institutions of higher learning do not address the changes needed, their colleges and departments of agriculture may eventually become irrelevant. Their graduates will have difficulty in keeping up with the changing needs of society and in securing stable careers. And the nation will miss its opportunity for leadership in addressing the global challenges related to food and agriculture.

RECOMMENDATIONS FOR CHANGE

The following recommendations for change are objectives, not prescriptions for specific actions. Across the nation, the institutions that house food and agriculture are very different from one another—they range from large research universities to two-year tribal colleges—so the notion of recommending that some particular new program or structural change be adopted by all of these institutions would be inappropriate and destined to fail. The committee believes that individual institutions must address each objective with interventions they develop, considering each institution's unique strengths, challenges, and circumstances. Although the full report provides examples of different approaches, drawn from those developed at different institutions, the most important aspect of the recommendations is the need for colleges and universities to commit to addressing these objectives. The final recommendation of the report calls attention to an appendix in the report, where a "checklist" of issues is contained. They provide the basis for self-evaluation that might provide institutions and others with a sense of how well they are making progress. Thus, the transformative power of the recommendations lies in the process of their implementation. The more of the objectives that are addressed, the greater their synergy, and the more positive their impact on teaching and learning and on the quality of the scholarship associated with colleges of agriculture in general.

RECOMMENDATION 1

Academic institutions offering undergraduate education in agriculture should engage in strategic planning to determine how they can best recruit, retain, and prepare the agriculture graduate of today and tomorrow. Conversations should involve a broad array of stakeholders with an interest in undergraduate agriculture education, including faculty in and outside agriculture colleges, current and former students, employers, disciplinary societies, commodity groups, local organizations focused on food and agriculture, and representatives of the public. Institutions should develop and implement a strategic plan within the next two years and to revisit that plan every three to five years thereafter.

Strategic planning should be the beginning of an extended and ongoing process of change, evaluation, and adaptation. Implementation will need to follow the ideas, pilot-testing, and continual assessment used to refine and improve new programs and policies. The committee emphasizes that action and implementation are necessary steps for achieving the goals of this recommendation and encourages academic institutions to include timelines for implementation as formal parts of their strategic plans.

RECOMMENDATION 2
Academic institutions should take steps to broaden the treatment of agriculture in the overall undergraduate curriculum. In particular, faculty in colleges of agriculture should work with colleagues throughout the institution to develop and teach joint introductory courses that serve multiple populations. Agriculture faculty should work with colleagues to incorporate agricultural examples and topics into courses throughout the institution.

Among the ways that more students can be exposed to agricultural topics are the incorporation of agriculture examples in courses outside agriculture and the offering of team-taught and interdepartmental introductory courses that serve students in a variety of majors. Agriculture colleges have a unique and continuing role if they can bridge the many academic domains that can contribute to a broader understanding of agricultural issues.

RECOMMENDATION 3
Academic institutions should broaden the undergraduate student experience so that it will integrate:

- **numerous opportunities to develop a variety of transferable skills, including communication, teamwork, and management;**
- **the opportunity to participate in undergraduate research;**
- **the opportunity to participate in outreach and extension;**
- **the opportunity to participate in internships and other programs that provide experiences beyond the institution; and**
- **exposure to international perspectives, including targeted learning-abroad programs and international perspectives in existing courses.**

During an undergraduate education, students should master a variety of transferable skills in addition to content knowledge. Employers value those skills at least as much as book learning. Providing students the opportunity to engage in a variety of experiences, such as those listed above, helps to make content knowledge come alive while strengthening the so-called soft skills important in the workplace. The ability to connect undergraduate education and extension is an opportunity unique to colleges of agriculture; it not only expands the sphere of institutional and statewide outreach but provides a chance for undergraduate students to give back to their communities and become spokespeople for agriculture.

RECOMMENDATION 4
Several actions are necessary to prepare faculty to teach in the most effective ways and to develop new courses and curricula:

- **Academic institutions, professional societies, and funding agencies should promote and support ongoing faculty-development activities at the institutional, local, regional, and national levels. Particular attention should be paid to preparing the next generation of faculty by providing appropriate training to graduate students and post-doctoral researchers. Moreover, academic institutions should take steps to ensure that the responsibility for faculty development rests not with individual faculty members but with departments, colleges, and institutions.**
- **Academic institutions and funding agencies should leverage existing resources or provide additional resources to support the development of new courses, curricula, and teaching materials. Among the needed resources are faculty release time, support for teaching assistants, attendance at education-focused workshops, and use of education materials and technologies.**

The scholarship of teaching and learning has developed substantially over the last several decades. Nevertheless, universities still tend to use an outmoded method of teaching in which lecturing is the norm and the focus on facts is predominant. Many classes fail to engage students or to take advantage of the research in how people learn. In general, university faculty do not receive much training in effective teaching, nor are they exposed to research in student learning; faculty in agriculture are no exception.

Therefore, it will be necessary for a variety of stakeholders to devote their attention to ensuring that current and future faculty members learn about the research on how people learn and have access to resources to implement course and curricular changes. The committee especially encourages graduate programs to build those topics and competences into training for the next generation of faculty.

Faculty will need access to professional-development opportunities and to the resources necessary for implementing effective instructional strategies. Educational innovation is generally much less expensive than investment in research, but it is not free. In fact, time may be a more precious resource than money for many faculty: time to develop new courses, redesign curricula, and identify, adapt, or create the necessary teaching materials.

RECOMMENDATION 5
Several stakeholders should take tangible steps to recognize and support exemplary undergraduate teaching and related activities:

- **Academic institutions should enhance institutional rewards for high-quality teaching, curriculum development, mentoring, and other efforts to improve student learning, including rigorous consideration in hiring, tenure, and promotion. Academic institutions should also implement new tenure-track faculty appointments that emphasize teaching and education research in a discipline.**
- **Funding agencies should support and reward excellence in teaching with education and research grants. Such models as the National Science Foundation's "broader-impacts criterion" should be considered by other agencies.**
- **Professional societies should raise the profile of teaching in the disciplines. That may include offering support and rewards for undergraduate teaching and sponsoring education sessions and speakers at society meetings, workshops on teaching and learning, education-focused articles in society publications, and efforts to facilitate the development and dissemination of teaching materials.**

Achievements in teaching are only rarely rewarded in substantive ways, so faculty are generally motivated to focus their attention elsewhere. That poses a particular challenge to the implementation of the recommendations in this report inasmuch as effecting change in undergraduate agriculture education will require attention to teaching and learning. Although a full vetting of tenure and promotion criteria and institutional priorities is well beyond the scope of this report, improving undergraduate education in agriculture depends on raising the profile of teaching.

RECOMMENDATION 6
Academic institutions offering teaching and learning opportunities in food and agriculture should enhance connections with each other to support and develop new opportunities and student pathways. In particular, four-year colleges and universities should further develop their connections with community colleges and with 1890 and 1994 land-grant institutions. In addition, four-year institutions should work with other institutions to establish and support joint programs and courses relevant to agriculture and develop pathways for students pursuing agricultural careers.

Academic programs in agriculture tend to exist in isolation, with few connections between institutions or even in the same geographic area. Community and tribal colleges are increasingly producing large numbers of students and especially high percentages of members of traditionally underrepresented groups for four-year colleges, but there are currently few pathways for those students to pursue agricultural careers. Articulation agreements and transfer partnerships should be developed between two- and four-year institutions when appropriate—but connections should not be limited to those arrangements. Institutions may wish to develop multi-institution programs, share resources, allow easy exchange of faculty and students, and generally work together to support and promote initiatives of common interest, regardless of an institution's official status as a land-grant institution.

RECOMMENDATION 7
Colleges and universities should reach out to elementary-school and secondary-school students and teachers to expose students to agricultural topics and generate interest in agricultural careers. Although the specific partnerships will differ from institution to institution, programs that might be considered include agriculture-based high schools, urban agricultural education programs, and summer high-school or youth enrichments programs in agriculture. In addition to formal partnerships and academic programs, colleges and universities should explore partnerships with youth-focused programs, such as 4-H, National FFA, and scouting programs.

The public perception of agriculture is a challenge beyond the scope of this report, but it is a factor that influences the perspective of future undergraduate students. Actions related to this issue cannot occur solely within institutions of higher education, but colleges and universities do have the capacity and responsibility to effect change in K–12 and other extracurricular programs. In fact, it is in the self-interest of these institutions to foster interest in and awareness of the role of agriculture in society among its youngest citizens.

RECOMMENDATION 8
Stakeholders in academe and other sectors should develop partnerships that will facilitate enhanced communication and coordination with respect to the education of students in food and agriculture. The partnerships should include the following elements:

- **Academic institutions should include representatives of industry and other employers on visiting committees, on advisory boards, and in strategic planning. Companies should include academic faculty on their advisory committees.**
- **Exchange programs should be developed that enable food and agriculture professionals to spend semesters teaching and working at academic institutions and enable faculty to spend sabbaticals working outside of academe.**
- **Opportunities for students to work in nonacademic settings should be developed and greatly expanded. Programs might include internships, cooperative education programs, summer opportunities, mentoring and career programs, job shadowing, and other experiences.**

There is a need to increase the permeability between academe and the private and public sector employers of graduates from agriculture programs. Industry has little understanding of how colleges and universities are organized, and academe has little understanding of industry and public sector needs. Although a number of universities have long-standing partnerships with particular industries or corporations, there are many opportunities to expand such collaborations to a wider array of private and public institutions, companies, and sectors. To reduce the "silo effect," the committee endorses steps such as those listed above that enhance communication and coordination between academe and employers of agricultural graduates at different levels.

Each of the elements in the recommendation is meant to provide a mutually beneficial relationship. For example, students benefit from such activities as internships and cooperative education programs to gain real-world work experiences, and industry gains an opportunity to recruit and attract talented young people and hire workers who already have experience working in the company. Closer connections between academe and industry may result in other opportunities, such as participation of the colleges in solving industrial challenges; such questions may serve as case studies in undergraduate classes and provide opportunities for undergraduate research.

RECOMMENDATION 9

Organizations and individuals conducting reviews related to undergraduate education in agriculture should incorporate the elements discussed in this report (summarized in Appendix E) to guide their decisions and reports. This includes accreditation, review of grant

proposals, department and other institutional reviews, and other venues.

In order to provide a strong incentive for implementation, the committee has developed a checklist of items that should be used by any individual or group conducting a review of a program, curriculum, department, college, or institution. The checklist includes questions about the nature of the curriculum, the ways that courses are taught, and the teaching style and knowledge of faculty about how students learn, among others. Although the committee does not have the authority to enforce specific competencies, it hopes that these elements will inform the establishment of review criteria and accreditation standards at all levels and in a wide variety of settings.

For example, the U.S. Department of Agriculture (USDA) might incorporate more specific elements into the evaluation criteria for the review of its programs as appropriate including—but not limited to—the Higher Education Challenge Grants Program. Accreditation bodies within the United States could use these elements to develop a specific set of benchmarks that institutions might be asked to meet to receive accreditation. External review and visiting committees might ask institutions and programs to meet the standards called for in this report. Peer-review panels might use the elements as goals that submitted grant proposals should seek to achieve. Professional societies could use these elements to guide discussions within disciplines and to make decisions of organizational priorities based upon those elements. The Association of Public and Land-grant Universities[1] can use the elements in this report to guide the content of teaching workshops and discussions among the members of its Academic Programs Section.

The committee hopes and expects that monitoring implementation and change will itself become a topic for research and evaluation. Faculty and graduate students in agricultural education programs may see this as a fruitful area for long-term study, tracking change and determining factors that contribute to institutional change and effective implementation.

CONCLUSION

In 1991, the National Research Council joined with the USDA to sponsor what was termed a landmark national conference to outline the changes necessary to meet the needs for professional education in agricul-

[1]Formerly known as the National Association of State Universities and Land-Grant Colleges.

ture. The present report considers the progress since that 1991 meeting and identifies opportunities to effect change in undergraduate programs that will enable programs to produce a flexible, well-prepared workforce.

In meeting its charge, the authoring committee engaged many people in academe, industry, professional societies, and interest groups, including those in the food and agriculture community and those outside the traditional group of stakeholders. Central to the committee's data-gathering was a Leadership Summit that brought together over 300 leaders, from undergraduate students to university presidents and from entry-level employees to CEOs of multinational food and agriculture companies. Discussions at the summit provided the committee with diverse viewpoints that were considering in drafting this report.

The committee recognized that undergraduate education in agriculture has changed fundamentally since 1991. The university and food and agriculture are different and have greater scope and scale. Teaching and learning have been informed by advances in how people learn and by a wealth of research on effective teaching methods and advances in instructional technology. We know better about what to teach and how to teach, but this knowledge is not always used to inform practice. Students are different—in background, in demographics, in interests, and in values. All those changes provide important background for the actions called for in the report.

Although conversations about improving teaching and learning in agriculture have been under way for many years, implementation has been slow. The time to act is now. The changes in our students, in our universities, in our society, and in our environment will not wait any longer. Only with a sustained commitment to improving education in agriculture will the necessary transformation occur. To maintain momentum, a continuous conversation will need to occur in universities and in disciplines, nationally across institutions and fields of inquiry. Agriculture departments, colleges, and institutions need to lead. The investment they make in undergraduate education will play a role in shaping the future of agriculture and its role in sustaining our world.

1

Motivating Change

Our world is changing at an increasing pace, producing many challenges that did not exist a generation ago. "Sustainability" is the watchword of today, linking issues from energy security to national security, from human health to the health of the planet. The challenges have intimate ties to food and agriculture, and colleges of agriculture[1] are in a perfect position to address them. With their mix of basic, applied, and social sciences, the colleges already have the fundamental—and historical—capacity to respond to complex issues, such as developing biologically based means of energy production, preserving the security and safety of our food supply, protecting the environment and using natural resources efficiently, and understanding the connections between nutrition and health to address important issues such as obesity.

If agriculture colleges are to lead the way to a future of continued well-being, they will need to recognize the key changes that are occurring and the influence of those changes on the skill sets needed by the next generation of leaders. The colleges will have to reform their undergraduate curricula and their students' experience to meet the needs of a changing world. This report discusses the practical meaning of that reform and outlines a path to effect the necessary changes.

[1]Throughout this report, the phrase *college of agriculture* and similar terms refer to administrative units that include food, agriculture, and related disciplines. In many cases, such a unit incorporates other disciplines, including natural resources, environmental science, and life sciences. The terms should be interpreted as including all such entities, whether colleges, divisions, departments, or other administrative units.

WHAT IS AGRICULTURE?

Agriculture can mean different things to different people. To some, it has been limited to production agriculture—that is, farming. While farming remains a vital and central part of agriculture, what defines 21st-century agriculture is much broader, encompassing a range of natural and social science disciplines. Uniting them is a commitment to understanding and sustainably and responsibly utilizing natural resources to benefit humanity. Therefore, it is the *motivation* behind the activity that defines something as part of agriculture. Agricultural disciplines can look quite similar to those traditionally outside of agriculture as researchers may pursue similar questions and use similar techniques. But agriculture can be distinguished by interest in the application of the work to agricultural systems, even when conducting basic research. Looked at in another way, agriculture often focuses on question of "how" in addition to "why": how to improve animal nutrition, how to grow crops without the use of pesticides, how to develop markets that support sustainable models for agriculture; in contrast, other disciplines tend to focus less on the "how" and are, instead, interested in understanding mechanisms or phenomena.

Throughout this report, the committee has taken an inclusive view of agriculture, including related disciplines that are sometimes considered separately. The reader should, therefore, take "agriculture" to include disciplines such as forestry and nutrition as well as related areas of natural resources, environmental science, and life sciences. In fact, one important message of this report is the degree of commonality between agriculture and related disciplines, meaning that an inclusive definition of agriculture is usually most appropriate.

THE NEED FOR CHANGE

To be sure, many institutions have made changes over the last several decades, and some of the ideas and best practices suggested here will be familiar to some readers. However, many students experience conditions that have not kept pace with the changing times. Even institutions that have been at the forefront of reform have not addressed all the challenges, so there are opportunities for *every* institution to discuss and improve. Moreover, institutions that have implemented many of the ideas discussed in the report can be leaders for those who are only now taking action. The committee hopes that all institutions will not only be receptive to the changes proposed, but also responsive.

Moreover, reform is not a one-time event. Academic institutions need to continually keep pace with the realities of 21st-century agriculture and agribusiness, and in particular, with the significant forces shaping agriculture today, including the integration of global agricultural markets; the growing concern for the environmental impact of agriculture; the scientific redefinition of agriculture; the effects of growing consumer influence; the push for local and organic foods; the need to respond to increasing rates of obesity; and the changing demographics of the agriculture workforce.

Global Integration

Scholars, pundits, and observers of all stripes regularly make the case that the world has changed and continues to change. Thomas Friedman (2005) suggests that the "world is flat"—in other words, that disparities in economic and creative opportunities across nations of the world are leveling out. Through social, cultural, political, and economic integration, we are now connected to one another in ways, both simple and complex, never before experienced.

In food and agriculture, competition for outputs (commodities and products) and inputs (fertilizer and fuel) reflects worldwide participation. America's farmers buy and sell in a global marketplace; indeed, agriculture is a cornerstone of U.S. trade activity. Markets, production, and distribution are both global and local, and knowledge about agricultural production is generated internationally and widely shared.

Public policy made in one country has implications well beyond national boundaries. What happens in Beijing, Jakarta, or Bogotá has ramifications in Minot, Austin, and Raleigh. A wide range of factors including currency exchange rates, distribution costs and capacity, environmental regulation, and labor cost differentials routinely affect the competitiveness of American agriculture. Challenges arise when those factors diverge widely between countries—for example when foods and raw ingredient sources from around the globe must meet safety and environmental standards that vary widely from one country to another. Increasingly, agricultural policy is shaped by multinational agreements and alliances.

Ultimately, there is a fundamental need to feed a growing world population. Addressing world hunger creates an imperative to provide healthful food worldwide. But the uneven availability of food, the difficulties in growing and transporting food, and the unpredictable nature of both humanitarian crises and natural disasters will further challenge the agricultural sector.

New Science

This is the era of "scientific agriculture." Genomics, ecology, chemistry, engineering, and other science disciplines play essential roles in 21st-century food and agriculture. As these disciplines become increasingly intertwined with food, fiber, and fuel production, agriculture has lost a little of its distinct identity. Agriculture now so thoroughly combines basic and applied aspects of the traditional STEM disciplines of science, technology, engineering, and mathematics that the acronym might rightly expand to become STEAM, joining agriculture with the other fundamental disciplines.[2] Agriculture can also connect with social science disciplines in areas such as ethnobotany and rural development, with medicine in areas such as pharmacognosy and nutrition, and with a large range of emerging and traditional fields from throughout the university.

Research and technology developed from public and private sources are primary inputs into agriculture and agribusiness. Despite a shrinking pool of researchers and declining support for research, the nation's colleges and universities, both within and outside of colleges of agriculture, are significant contributors to the scientific basis of the agricultural enterprise.

Consumer Influence

Consumers in the United States are increasingly interested in all aspects of the food they eat. They expect abundant, affordable, safe, and healthy foods and want a wide array of food products and choices year-round. But they increasingly look for humanely produced and environmentally sound products that are "organic," "natural," and "local." Americans are expressing their demand and expectation of agricultural producers through the market, as evidenced by the remarkable growth of farmers markets, community-supported agriculture, agrotourism, and the emergence of "slow food" groups. Consumers have also exercised their influence through public policy measures that, for example, proscribe certain types of plant and animal production or subsidize school purchases of locally grown food.

Consumers also demand nonfood products from agriculture, such as natural fibers for clothing and textiles. Nursery products, ornamentals, and turf grass have become important growth industries and, of course, forestry

[2]Since this report was issued in prepublication form, the committee has learned that this use of STEAM education was independently coined by Dr. John Nishio, Director of the Professional Science Master's Program in Environmental Sciences, at California State University, Chico.

and lumber products are also agricultural products. More recently, consumers and other sectors are turning to agriculture to produce fuels and energy products.

Even as science and engineering advance to better serve agriculture, some segments of the public remain skeptical about the putative benefits of scientific advances. Finding a way to reconcile the potentially conflicting demands of consumers will be a challenge to agriculture in the years ahead.

Environmental Concerns

While responding to multiple new demands and expectations, American agriculture is increasingly concerned about environmental damage and natural-resources sustainability. How will professionals in agriculture use water and manage soils and land responsibly? How will they protect fresh-water supplies and maintain air quality? How can agricultural materials be ethically sourced? Moreover, the effect of climate change on food and agriculture constitutes an important unknown for the future of our food system and production agriculture.

American agriculture will be fundamentally influenced by the rapidly emerging challenges of providing, allocating, managing, and conserving water and energy. Some farmers are becoming part of the alternative-energy production system, and others are being adversely affected by the runup in energy prices. Some have the capacity to adapt to the new realities of scarce water, and others will probably face serious consequences.

The dynamics of energy and water will ultimately restructure agriculture substantially and will redefine agriculture's relationships with the larger economy.

Demographic and Political Shifts

The traditional "farm population" now makes up less than 2% of the U.S. population. Fewer citizens than ever before now play a role in agriculture, and public understanding of what is involved in the food and fiber system has decreased. One result is that a once powerful farm lobby is losing clout, particularly in the federal policy-making process. An increasing number of voices now have a stake in agriculture policy, not only the traditional agriculture-based organizations and not only those with a high degree of agricultural literacy.

The days when agriculture-related employers could expect to hire new employees with farm backgrounds are over. There are not enough "farm kids" available. Even the land-grant institutions in farm states are largely and increasingly populated by students with urban and suburban backgrounds. These students, who come from a diversity of cultural, economic, and ethnic backgrounds, bring a variety of ideas and skills to the agricultural enterprise—and changing expectations for their undergraduate education.

IMPACT OF CHANGES ON AGRICULTURAL EDUCATION[3]

As a consequence of the many changes in agriculture and related industries, employers seek growing sets of skills and perspectives in the people they hire. Clearly, people with global perspectives and concern for the environment increasingly will be in demand, as will those with rigorous scientific preparation in a variety of fields. But other skills are also essential, including problem-solving, critical thinking, team-building, leadership, communication, conflict and financial management, and thriving in diverse environments. Thus, the agriculture-related sectors seek employees, managers, and leaders who bring a wide variety of skills with an appreciation of what agriculture is today.

Industry leaders and other employers look to colleges and universities to produce employment-ready graduates who meet the new and emerging standards. They will hire qualified students wherever they are. Companies that have traditionally hired graduates from colleges of agriculture are increasingly looking elsewhere in the university. They are finding equally—or even better—qualified students in colleges of arts and sciences, colleges of engineering, and throughout the university.

Even as agriculture confronts powerful new forces and the accompanying challenges, agriculture remains essential to America's economy and the way of life of many people. Agriculture, of course, produces the essentials of life, but it also constitutes a major national economic sector and a primary player in both international and local commerce.

Much of rural America continues to depend on agriculture and agribusiness as drivers of economic development and social stability. As the stewards of natural resources, agricultural leaders will continue to play a central role in the long-term strength of local communities.

Maintaining a strong and vibrant agriculture system is central to national security and economic competitiveness. Innovation and resource allocations

---

[3]Throughout the report, the term *agricultural education* is used to refer to undergraduate education in food, agriculture, and related disciplines. It is not meant to refer specifically to the discipline of agriculture education.

must be brought to bear at every level if agriculture is to master the challenges of the future.

THE ROLES OF LAND-GRANT UNIVERSITIES AND OTHER INSTITUTIONS

The advance of land-grant universities, arising from the Morrill Act of 1862 and additions in 1890 and 1994, has been a profound and responsive innovation. Several acts of Congress added to the original teaching mission of land-grant institutions, but education remains a central component in the social contract of land-grant universities. Although necessary to cast it in a contemporary context, the directive in the Morrill Act to "promote the liberal and practical education of the industrial classes in the several pursuits and professions in life" remains relevant today in all land-grant institutions.

Of course, agricultural education is not limited to land-grant universities. A large number of other public and private institutions of higher education offer instruction in food and agriculture and should be seen as among the prime audiences for this report.

Just as agriculture will need to adapt to progress, colleges and universities will have to change to advance education and scholarship in agriculture, agribusiness, and natural resources effectively and to foster enhanced public literacy about these issues.

Colleges and universities, including land-grant institutions, should produce employees, managers, leaders, policy-makers, and natural and social scientists who accept and respond to the dynamic world of agriculture and agribusiness.

CONTINUING PROMISE OF THE AGRICULTURAL EDUCATION AND THE LAND-GRANT SYSTEM

With all the changes taking place in the world and in academe, one might ask whether agricultural education and the land-grant university system are relics of the past. Has the agricultural mission of land-grant universities outlived its usefulness?

The committee emphatically answers no to those questions. Food and agriculture offer many opportunities for the future, and contributions of these disciplines are essential for addressing some of the most difficult societal challenges. This report strongly calls for reinvigoration of undergraduate education in agriculture and a reaffirmation of the land-grant university and of undergraduate education in agriculture. Fewer students will be directly engaged in farming, but there will still be a great need for citizens who have a deep understanding of the agriculture system. In fact, with increasing globalization, advances in technology, and the need for the public to make decisions about agricultural issues, the need is stronger than ever. Agriculture is linked not only to such traditional sectors as food and textiles but increasingly to such 21st-century challenges as energy production and the protection of our environment.

Agriculture and the land-grant university system are well positioned to take advantage of what today's students are demanding, as they have done for years. Perhaps one of the most important things agriculture needs to do is "rebrand" itself. For example, land-grant institutions were set up to respond to the needs of the day, meaning that such institutions have a responsibility to adapt to changing times. They have a compelling reason to communicate to the public and to students that agriculture not only is not behind the times but it also has the necessary qualities to lead the way into the future:

- Agriculture colleges incorporate outreach at their core; in fact, they often have the most extensive extension activities and may even be the *only* part of universities that have an explicit responsibility to reach beyond the walls of the institution and engage the public. What better way to appeal to students who want to make a difference in the world and work toward affecting the lives of others directly and beneficially than through this existing structure and the network of extension centers?
- Agriculture focuses on outcomes and results. Although many agricultural scientists conduct basic research on plant, animal, and microbial systems, there is a strong emphasis on application. Investigation is often motivated by a desire to realize specific objectives. For that reason, many scientists focus on solving specific challenges or moving a system in a particular direction. Such a results-driven mission allows both students and scholars to work directly on problems that have important implications for the well-being of society.
- Agriculture and the disciplines that make up agriculture colleges bring basic and applied sciences together. Outside the agriculture college, there is often a tension between those conducting basic research and those applying its results to develop products and applications. But agriculture colleges themselves integrate science and practice in the same research projects.
- Agriculture colleges often include biological-, physical-, and social-science departments within the same college. That provides a unique opportunity for interdisciplinary research and teaching that can serve as a model for the university.

• Agriculture integrates the laboratory and the field. Many scientists either work in a laboratory or go out into the field. Agriculture includes both, by conducting laboratory investigations and exploring what will be most effective in the field.

• Agriculture is intricately intertwined with various aspects of environment and natural resources. In fact, there is little in agriculture practice that does not have important connections with the environment. Therefore, students and scholars interested in environmental stewardship will find many opportunities for working on these challenges in agriculture.

• Agriculture is also intertwined with all aspects of food production and nutrition. Addressing such challenges as hunger, obesity, and nutrition will require professionals with a firm grounding in the agricultural sciences.

THE CONSEQUENCES OF FAILURE

Failure to respond to the changes affecting agriculture and education will place many aspects of the nation's universities, agriculture system, and society at risk. The agricultural community—by whom this report is written and to whom it is addressed—has a responsibility to ensure that agricultural education is appropriate for changing times. Failure could put agriculture itself at risk. With the impending retirement of the "baby boom" generation, rebuilding the human-resource base of agriculture will be critical to its future. Failure could mean denying many the opportunity for a career in an exciting and rewarding industry. Failure could mean the decline and marginalization of our colleges and universities themselves. Failure could mean that the United States will fall behind other nations in agriculture-based science and stewardship. And failure could contribute to the loss or pollution of our land, water, and natural resources.

GOALS OF THE REPORT

This study emerged from conversations with the National Association of State Universities and Land-Grant Colleges (NASULGC)[4] and its Academic Programs Section. After receipt of sponsorship from government agencies and private foundations and organizations, a National Research Council committee was convened to consider the changes needed in undergraduate agricultural education to produce a flexible, well-prepared workforce that is appropriately skilled, socially responsive, and technically proficient (see

[4]NASULGC is now known as the Association of Public and Land-grant Universities.

Appendix A for the complete statement of task). This report seeks to chart a course for agriculture graduates to be prepared for a wide range of careers in food and agriculture—whether they work in fields or laboratories, boardrooms or courtrooms. While farming remains an appealing career for many students and those with expertise in production agriculture are needed, the range of career options in food and agriculture is much broader than it was a generation ago.

As part of the study, the committee and project staff organized a seminal event to draw attention to the need for change in undergraduate education in agriculture. The event, the Leadership Summit to Effect Change in Teaching and Learning, drew over 300 people from academic institutions, business and industry, government agencies, professional societies, and other stakeholders. The presentations and discussion at the Leadership Summit and other speakers and input to the committee from diverse sources helped to provide context and varied perspectives and allowed the committee to consider the issues broadly.

The committee saw its role as recommending a structure for change, allowing institutions and the agriculture community to adapt to continually changing times. Although recruiting, retaining, and graduating the best students and providing them with the skills to succeed in future careers is at the heart of the report, the main thrust of the report is in establishing the structures that will make this happen. Time and time again through the study process, there was a clear message that institutions need to be "nimble" and be able to adjust to new circumstances and take advantage of arising opportunities. In part for this reason, the committee has chosen not to make overly specific recommendations with detailed curricula or precise programs since those ideas would necessarily be out of date within a few years. Just as this report argues for preparing students to learn and adapt, the report calls upon institutions to do the same. This is what will sustain institutions and ensure that the education they offer remains relevant.

The committee believes that it is important for the report to be not only visionary, but practical and possible. Since spurring action is one of the main goals of the report, the proposed changes must be realistic and actionable. So, for example, the committee could have recommended that agriculture colleges be disbanded and their constituent departments folded into the various other colleges at their institutions. But, whether or not dissolving colleges of agriculture is the right course of action, the committee decided that it was unlikely to happen. The committee could have called upon state legislators, members of Congress, and officials at federal agencies to enact substantial increases in funding for universities and for undergraduate education. As welcome as these additional resources might be, the committee was realistic that such additional resources might not be easily forthcoming, and change must be implemented even without them. The committee could have outlined a specific series of courses for each of the several dozen majors that might be offered by a college of agriculture. Yet this would unnecessarily constrain the ability of institutions to make priorities on their own strengths and areas of expertise—and would have been out of date as soon as the report was printed.

The committee encourages institutions to consider the messages and recommendations seriously, to ask how they can best achieve the goals highlighted in the report, and to anticipate the results of reaching beyond the status quo; institutions should also consider the consequences of *inaction* as a decision not to change is an action nonetheless. Therefore, the focus is on *how* to bring about change and on the structures and policies that will enable institutions to provide the best undergraduate experience and to recruit and retain the best students for the careers of today and tomorrow.

ORGANIZATION OF THE REPORT

Chapter 2 provides additional context and background underlying the changing nature of undergraduate education in agriculture. Chapter 3 summarizes the research and opportunities to reform teaching and learning. Chapter 4 discusses the need for breaking down silos within the university, focusing on interdepartmental and cross-college collaboration. Chapter 5 highlights opportunities for partnerships that extend the reach of the university to other types of institutions and organizations. Finally, Chapter 6 outlines the steps that are needed by compiling the committee's conclusions and recommendations.

Several appendixes are also included: the committee's statement of task; information about the October 2006 Leadership Summit, including two background papers; a checklist for the review of programs and institutions; and biographical information about the committee and staff.

2

The Context for Change

Urgent change is required in agricultural education. To be sure, change is already occurring—and has been for a number of years—but there is a need for action in particular directions. The change needed today is a refocusing on the undergraduate curriculum and student experience so that the agriculture graduates of tomorrow will have the skills and competences to meet the needs of a changing workplace and world.

CHANGE IN STUDENTS

Students of the 21st century differ from those of the last century in many ways, including a demographic change: fewer come from farm or rural backgrounds. Today, well under 5% of the U.S. population live on farms, and barely 20% come from rural communities (Dimitri et al. 2005). The increasingly urban and suburban population poses a particular challenge for agriculture in that students often lack even basic awareness of agricultural sciences. For example, a 2006 survey of academic program administrators found "misconception or image about the agricultural sciences" was the most important concern affecting the selection of agricultural sciences as a career by U.S. high-school students (Gonzalez 2006).[1]

Public understanding of agriculture is poor, and many people are barely aware of where their food comes from. Their lack of awareness of agricultural products is coupled with an outdated view of agriculture. One challenge for attracting undergraduate majors to agriculture is therefore to

[1]The other factors, in decreasing order, revealed by the same survey were lack of knowledge about employment opportunities, lack of knowledge about fields of study, perceived relevance to or importance for future careers, lack of fundamental knowledge of mathematics and sciences, and peer and family pressure against agricultural-science studies.

overcome the public perception that agriculture means farming, even though agriculture incorporates a wide array of questions and approaches.

Even as the number of college students in all fields of study has increased over the last several decades, the number of students earning degrees in agriculture has been relatively stable since 2000 (Figure 2-1). According to a background paper prepared by Gilmore et al. (2006) for participants in the Leadership Summit (see Figure C-1 in Appendix C), much of the growth in the number of baccalaureate degrees in agriculture and natural resources can be attributed to a small number of disciplines. For example, baccalaureate degrees in natural-resources conservation and research increased by a factor of about 5 between the 1987–1988 and 2003–2004 academic years. Degrees in agricultural business and management increased by about 15%, and in animal sciences by more than 25% in the same period.

Much of the growth in baccalaureate degrees can be attributed to the increase in the number of women pursuing undergraduate study in agriculture and natural resources. Men earned almost twice as many agriculture bachelor's degrees as women in 1987–1988, but near parity between

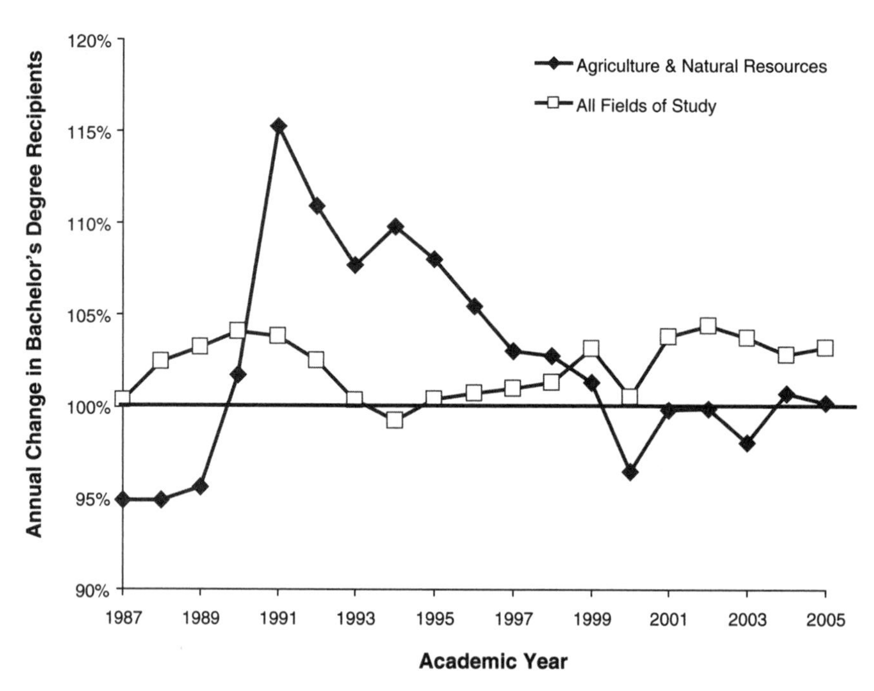

FIGURE 2-1 Annual change in bachelor's degree recipients for agriculture and natural resources and all fields of study, 1987–2004.
SOURCE: National Center for Education Statistics (Snyder et al. 2009).

the sexes is observed in data from 2003–2004. Moreover, the number of agriculture bachelor's degrees earned by men has been decreasing since 1995–1996.

Despite progress toward gender equity, there has been relatively little progress in broadening the participation of underrepresented minorities in agriculture. The percentage of Black/African-American, American Indian and Alaska native, Asian and Pacific islander, and Hispanic baccalaureate-degree recipients has increased only modestly over nearly a decade (Gilmore et al. 2006). The number of Hispanic graduates now exceeds the number of Black, non-Hispanic, and Asian and Pacific islander graduates. Racial and ethnic diversity is particularly important for the future of agriculture as the percentage of members of underrepresented groups increases in the United States. For example, underrepresented minorities made up nearly 40% of K–12 students in 2002; this suggests that the undergraduate population of the next generation will be much more diverse than that of the past.

Student interests and motivations are also changing. Students want careers that are going to provide a steady income, but they also want to pursue careers that will be personally and professionally rewarding, provide an appropriate work–life balance, have the image of a 21st-century professional, and are aligned with their values and interests. Agriculture faces a particular challenge in this regard because careers in agriculture may appear to be outdated, may not pay top salaries, or may not be perceived as offering sufficient opportunities for creativity. In addition, some fields in agriculture may be seen as in conflict with students' values in, for example, environmental stewardship and responsible land management. Many of these claims have to do with appearance, not substance, but it will be incumbent on the agriculture community to find mechanisms for attracting the most talented students to degrees and careers in agriculture.

There is no single kind of student, no single type of institution, and no simple set of solutions (Taylor 2008). In considering the recommendations and ideas in this report, stakeholders will need to consider the needs of different populations of students and of individuals. What are the implications of each change for agriculture majors? Nonmajors? What is the effect of lifelong learning? Of agricultural literacy in the entire population?

CHANGE IN INSTITUTIONS

The land-grant university system was established by the Morrill Act of 1862 and signed into law by President Abraham Lincoln. The act donated public land to the states and territories to establish "colleges for the benefit of agriculture and the mechanic arts." The act specifically called for the establishment of

> at least one college where the leading object shall be, without excluding other scientific and classical studies, and including military tactics, to teach such branches of learning as are related to agriculture and the mechanical arts, in such manner as the legislatures of the States may respectively prescribe, in order to promote the liberal and practical education of the industrial classes in the several pursuits and professions in life [7 U.S.C. 304].

The Second Morrill Act of 1890 expanded the pool of land-grant institutions to include institutions that enrolled Black students (7 U.S.C. 323). And 29 tribal colleges and universities were given land-grant status under the Equity in Educational Land-Grant Status Act of 1994.

Since the establishment of the land-grant university system nearly 150 years ago, the role of higher education in general—and the land-grant system in particular—has changed dramatically. What were once institutions focused primarily on agriculture and the mechanical arts have evolved to become world-class universities in which agriculture may be only a small part of the mission.

The officially designated land-grant universities are supplemented by a large number of other public and private institutions that offer instruction in food and agriculture. Sometimes referred to as non-land-grant colleges of agriculture, these institutions produce a sizeable percentage of the undergraduate degrees in agriculture. The committee intends to include these colleges and universities in all of its conclusions and recommendations.

A survey of colleges with *agriculture* in their names yields an interesting picture of institutional change. Among the states, Rhode Island and Massachusetts no longer have a college with *agriculture* anywhere in its name.[2] In the remaining 48 states, 67 colleges have *agriculture* in their names, but only about one-third are titled just College of Agriculture. The other two-thirds have additional names in their titles, clearly identifying an expanded mission. The most common combinations include *agriculture* with *natural resources*, *life sciences*, *environment*, or *food science*. Although focusing on titles may seem trivial, it does show that these colleges have

---

[2]Agriculture-associated disciplines are found mostly in the College of Environmental and Life Sciences at the University of Rhode Island. Massachusetts has two land-grant institutions: the Massachusetts Institute of Technology and the University of Massachusetts Amherst; agriculture itself is found in the latter, in the College of Natural Resources and Environment.

embraced a broader mission than traditional production agriculture, and in fact many are providing research and teaching in a variety of basic and applied disciplines.

Given the variety of names and missions, it is worth considering which qualities are shared by all colleges of agriculture. Although the first Morrill Act of 1862 was created to support education in agriculture and mechanical arts, it was recognized from the start that the colleges had to do more and not exclude other scientific and classical studies. In 1887, Justin Smith Morrill, the man behind the act, said the following at the Massachusetts Agricultural College (Morrill 1887, p. 20):

> It would be a mistake to suppose it was intended that every student should become either a farmer or a mechanic when the design comprehended not only instruction for those who may hold the plow or follow a trade, but such instruction as any person might need—with "the world all before them where to choose"—and without the exclusion of those who might prefer to adhere to the classics.

The inclusive role of land-grant institutions in providing education in a range of disciplines has meant that many such institutions are among the nation's premiere institutions in many areas. In fact, in many land-grant universities, colleges of agriculture often receive far less attention, more limited resources, and fewer students than colleges of law, medicine, business, engineering, and the liberal arts and sciences.

The federal investment in higher education in agriculture has also gone through a transformation as the U.S. Department of Agriculture (USDA) has expanded its support of higher education. USDA now invests more than $100 million a year through 20 national initiatives that support agricultural and natural-resources colleges both in and outside the land-grant system. It should be noted that although USDA investments have helped many colleges to update their curricula, facilities, and teaching methods, the amount of resources dedicated to instruction pales in comparison to federal funds allocated to research and extension. In a time of constrained state budgets, which play a critical role in supporting many institutions that offer instruction in agriculture, relatively small amounts of funds from federal agencies and private sponsors would be especially valuable. Moreover, federal requirements for institutional cost sharing in a number of programs, including graduate student fellowships, further constrain the ability of institutions to dedicate resources to undergraduate education. The committee hopes that support available from USDA will be supplemented by resources from other government agencies, institutions themselves, and other stakeholders to continue—and accelerate—the process of reform. Extramural support can serve as an important motivator. Even limited investment or subtle changes to program descriptions and review criteria to promote change in curricula and teaching methods can have a powerful impact nationwide.

In addition to the changes in students described above, many faculty in colleges of agriculture have different backgrounds and experiences from faculty of the past. Like their students, faculty are less likely to come from agrarian backgrounds or to have life experiences on the farm. Teaching and research have shifted from production practices to basic natural and social sciences and can sometimes be hard to distinguish from research and teaching in colleges of medicine or arts and sciences.

CHANGE IN HIGHER EDUCATION

The land-grant university system operates in the context of American higher education, which is increasingly concerned with accountability and efficiency. For example, former U.S. Secretary of Education Margaret Spellings convened a Commission on the Future of Higher Education in 2005, whose final report highlighted the need for "improved accountability" and increased transparency about student success; it also recommended the development of "new pedagogies, curricula and technologies to improve learning, particularly in the areas of science and mathematics" (U.S. Department of Education 2006).

The Voluntary System of Accountability (VSA) has been developed by the American Association of State Colleges and Universities and the Association of Public and Land-grant Universities to demonstrate accountability and stewardship to the public, to identify effective educational practices by measuring educational outcomes, and to compile information to facilitate comparisons among institutions.[3] The VSA will provide consistent portraits of higher-education institutions—including information about student engagement and core educational outcomes—that will be helpful to students, institutions, policy-makers, and other interested stakeholders.

Higher education was once available to only a small number of people, but it is becoming more common and even necessary for all students to pursue postsecondary education. In many cases, higher education is being pursued at expanded state universities, where undergraduate enrollment can measure in the tens of thousands. But postsecondary education is also occurring in greater numbers at a wider array of institutions, including com-

[3]See <http://www.voluntarysystem.org/> for more information about the VSA.

munity colleges, for-profit degree-granting institutions, and online universities. All those changes have the potential to fundamentally alter the role of and opportunities afforded by land-grant universities.

CHANGE IN AGRICULTURE

As discussed in Chapter 1, agriculture of today is not the same as it was a decade or a generation ago—and it is critical that agricultural graduates are prepared to meet the changing times. The disciplines that make up agriculture have changed to incorporate new ideas from the natural and social sciences and are sometimes hard to distinguish from similar departments elsewhere in the university.

Students will need to appreciate the systems nature of agriculture, gaining exposure to the breadth of agriculture and having opportunities to integrate what they learn in different courses. The systems approach incorporates not only the disciplines that traditionally comprise agriculture colleges, but other fields of study throughout the university. Agriculture now asks questions that cannot be confined to a single discipline: What is the effect of a given practice on the environment? What resources will be needed for a plan to be completed? What is the nutritional effect of a particular genetic modification?

Agriculture, like other sectors, operates increasingly across international boundaries, with even fresh fruits and vegetables shipped around the world; this introduces a complex regulatory regime, transportation logistics, and the need to work with different cultures, laws, and individuals. This intertwining of agriculture, culture, regulations, and concerns makes critical the need for professionals who have international exposure and sensitivities. As increased demand for resources is met with international supply bases and more domestically produced products are sold overseas, food and fiber professionals will need to understand the global implications of their research, their product designs, their market plans, or their individual growth potential. Having international experiences early in their training should broaden the scope of students' curiosity and prepare them for future work in an international marketplace.

At the same time as it is becoming international, agriculture is also becoming local. With a greater focus on locally sourced goods and the interconnection of agriculture with the development of rural communities and environmental stewardships, agriculture graduates will need to appreciate the ways that agriculture interacts with local environments.

CHANGE IN CAREERS

A USDA analysis of employment opportunities (Goecker et al. 2005) summarized by Gilmore et al. (2006) predicts a decrease in the number of positions available to undergraduate and graduate students with training in agriculture and natural resources. Strong employment growth in many management and business occupations is predicted, especially in such careers as technical sales, accounting and financial management, market analysis, landscape management, and international business. In contrast, weaker opportunities for those who provide services to farmers and ranchers are expected.

In scientific and engineering occupations, the analysis predicts growth in fields that take advantage of modern scientific advances, such as genomics, bioinformatics, breeding, biomaterials engineering, nanotechnology, and environmental sciences (Goecker et al. 2005). Fewer opportunities are expected in agricultural machinery, wildlife science, and veterinary sciences.[4] In agriculture and forestry, growth may be expected in specialty crops and materials that have use in medical or energy applications, landscape planting and trees, turf production, and aquaculture and organic farms. However, opportunities for producers of traditional commodities (such as wheat, corn, cotton, soybeans, cattle, and hogs) will continue to decrease. Finally, the USDA analysis predicts increasing opportunities in plant and animal inspection, public-health administration, nutrition, and environmental planning.

Employers of today are emphasizing skill development, not only content knowledge. For example, a study conducted by the National Food and Agribusiness Management Education Commission asked agribusiness employers to identify the most important skills, capabilities, and experiences needed by new college graduates. Topping the list were transferrable competences, including interpersonal communication skills, critical-thinking skills, writing skills, and computing skills (Boland and Akridge 2006; see Table C-3 in Appendix C).

Academic institutions will need to alter the focus of their academic programs and the experiences that they offer to students to keep up with the changing careers and opportunities available to their graduates.

---

[4]A National Research Council report of an assessment of the current and future workforce in veterinary medicine is expected to be completed in 2009.

IMPLEMENTING CHANGE

The committee encourages institutions to engage in serious consideration of and contemplation about the issues discussed and recommendations offered in this report. It will not be possible for every academic institution to implement every idea recommended here, and it will be necessary for universities and other stakeholders to set priorities for their actions. For that reason, some of the reaction to this report may be a choice *not* to do particular things or possibly even to eliminate or consolidate existing programs. Targeted excellence may be preferable to universal mediocrity. However, stakeholders should consider how to be sure that every student has the opportunity to take advantage of a suite of experiences, whether or not they can be offered by a given institution.

3

Improving the Learning Experience

Changes are needed in the undergraduate experience in agriculture. The changes include new curricula and content, but it will also be vital to improve how learning and teaching occur. This chapter will describe aspects of teaching and learning that are in need of reform, with a focus on the disciplines within food and agriculture. It therefore serves as context and background for readers who may not be familiar with the research on teaching and learning. The committee hopes that implementation of these ideas will help to both enhance the relevance of undergraduate education and retain students in agriculture. The chapter also provides a number of examples of research-based teaching strategies and discusses ways to raise the profile and impact of high-quality teaching within institutions and disciplines. Interested readers are encouraged to consult some of the many excellent reports on undergraduate teaching and learning that have been published in the last several years (e.g., AAAS 2004; Boyer Commission 1998; NRC 1996a, 1997, 1999c, 2003abc; Seymour and Hewitt 2000; Tobias 1992).

As throughout most of higher education, teaching in agriculture is strongly influenced by the skills and motivation of the faculty. Most teaching is good, but all teaching can be improved.

Effective teaching in higher education incorporates pedagogical strategies that create hospitable classroom climates supporting diverse learning processes and cultural understanding. The traditional approach to college-level instruction—especially in science, technology, engineering, agriculture, and mathematics disciplines—has historically been lecture-based delivery; as discussed below, the passive lecture format may not be as effective as desired in promoting student learning. Tutorials, laboratories, field-based learning experiences, problem-based learning, and other models can be especially effective in reaching students.[1] It should be noted, however, that hands-on activities are not always "minds-on." Effective educational activities require planning and structure to support student learning and achieve learning objectives.

Most higher-education faculty members arrive at their teaching positions after earning research doctorates. Few receive any formal training in how to be effective teachers or are exposed to pedagogy, the science of teaching. In fact, when thrust into the classroom, most faculty members teach the way they were taught during their own student experiences—which, for most, is almost exclusively lecture-based—despite research demonstrating that interactive engagement is more effective in enhancing student learning. It is not that faculty are unwilling to use research-based methods; rather, few have had the opportunity to expand their repertoire with new teaching techniques and tools or to learn about the need to provide more student-centered learning environments.

The committee applauds the work of a number of professional societies and journals that are committed to agricultural and undergraduate education (several of which are listed in Box 3-1). In drawing attention to the challenges and opportunities in agricultural education, the committee hopes that the community will call upon these organizations, joining as members, attending their meetings, publishing in their journals, and benefiting from their many years of scholarship.

Since the 1991 conference on undergraduate education in agriculture (compiled in NRC 1992), there have been important advances in the science of learning. The National Research Council volumes titled *How People Learn* (NRC 1999ab, 2005b) provide an excellent summary of what has been learned from education, cognitive science, psychology, and related fields and how to apply it to classroom practice (see Box 3-2).

The National Research Council's 2003 report on undergraduate teaching in the STEM disciplines (NRC 2003b)—science, technology, engineering, and mathematics—provides an excellent overview of the concepts that influence learning and synthesizes them into seven principles that may be useful for universities in thinking about reforming classes and curricula:

- Learning for understanding is facilitated when new knowledge and existing knowledge are structured around the major concepts and principles of the discipline.

---

[1]For a recent discussion of student laboratory experiences with application to both undergraduate and high-school laboratories, see NRC (2005a).

BOX 3-1
Selected Resources for
Undergraduate Education in Agriculture

Professional societies and associations:

- American Association for Agricultural Education: <http://aaaeonline.org/>
- Association for International Agricultural and Extension Education: <http://www.aiaee.org/>
- National Association of Agricultural Educators: <http://www.naae.org/>
- The National Council for Agricultural Education: <http://www.teamaged.org/>
- National Farm & Ranch Business Management Education Association: <http://www.nfrbmea.org/>
- National Postsecondary Agricultural Student Organization: <http://www.nationalpas.org/>
- National Young Farmer Educational Association: <http://www.nyfea.org/>
- North American Colleges and Teachers of Agriculture: <http://www.nactateachers.org/>

In addition, many disciplinary societies have sections and committees dedicated to issues of education, a number of whom have developed extensive resources and programs.

Journals:

- *CBE–Life Sciences Education*: <http://www.lifescied.org/>
- *Community College Journal of Research and Practice*
- *Journal of Agricultural Education*: <http://aaaeonline.org/jae.php>
- *Journal of Career and Technical Education*: <http://scholar.lib.vt.edu/ejournals/JCTE/>
- *Journal of Extension*: <http://www.joe.org/>
- *Journal of International Agricultural and Extension Education*: <http://www.aiaee.org/journal.html>
- *Journal of Natural Resources and Life Science Education*: <http://www.jnrlse.org/>
- *NACTA Journal*: <http://www.nactateachers.org/nacjournal.htm>

• Learners use what they already know to construct new understanding.

• Learning is facilitated by the use of metacognitive strategies that identify, monitor, and regulate cognitive processes.

• Learners have different strategies, approaches, patterns of abilities, and learning styles that are a function of the interaction between their heredity and their prior experiences.

BOX 3-2
How People Learn

The National Research Council reports *How People Learn* (NRC 1999ab, 2005b) reveal several principles that can help to guide instruction:

- New knowledge is built on a foundation of existing knowledge and experience. Everyday conceptions are resilient and must be actively challenged and engaged to support conceptual change.
- Learning for understanding requires a deep foundation of knowledge, understanding facts and ideas in the context of a conceptual framework, and organizing knowledge for effective retrieval and use.
- Metacognitive strategies help students to learn and take control of their own learning. These strategies—such as predicting outcomes, explaining to oneself, and noting failures of comprehension—can be taught effectively in the context of subject matter.

Those principles suggest that it is important for faculty to know the common conceptions and misconceptions that students bring to a topic, to directly engage students in confronting those conceptions, to use formative assessment to monitor student thinking, and to adapt teaching based upon the assessment. They also show the importance of determining the core concepts that organize a discipline, of structuring topics to support conceptual understanding, and of paying explicit attention to reflective assessment.

Summit presentation: M. Suzanne Donovan, *Program Director, Strategic Education Research Partnership Institute; Study Director,* How People Learn, *National Research Council*

- Learners' motivation to learn and their sense of self affect what is learned, how much is learned, and how much effort will be put into the learning process.
- The practices and activities in which people engage while learning shape what is learned.
- Learning is enhanced by socially supported interactions.

Once curricula are designed and implemented, the path to education reform is far from complete. One especially important component is assessment, which enables an instructor or an institution to determine whether a particular activity, course, or major has been effective in meeting its learning goals. How else is it possible to determine whether students have learned what is being taught? NRC (2003b) provides an overview of research on effective assessment of student learning that may be helpful to institutions and faculty members. The document provides a wealth of information beyond the several points mentioned here, but some elements bear highlighting:

- Multiple assessment measures provide a more robust picture of what a person has learned.
- Educational assessment must be aligned with curriculum and instruction if it is to support learning.
- Assessment practices should extend beyond emphasis on skills and discrete bits of knowledge to encompass more complex aspects of student achievement.
- Assessment should provide timely and informative feedback to students on their learning to inform the practice of a skill and influence effective and efficient acquisition.
- Assessment must be designed from the beginning of the instructional process to ensure that the desired type of information is available for assessment.
- For assessment to be effective, students must understand and share the goals for learning that are assessed.

The scholarship of teaching and learning has emerged as an important area of focus in higher education and has itself become a subject of research, following the impetus provided by Boyer (1990). That has led to academic study of and research on the most effective methods of teaching and even to the creation of subdisciplines on discipline-based education research in several fields. But the research base can help in the improvement of teaching and learning only if mechanisms that facilitate faculty implementation of the results of the research are put into place.

Effective teaching and learning have become even more important as the student body has become more diverse, but effective teaching has benefits for *all* students. Students have different learning styles and different ways of assimilating information, and using a "one size fits all" approach in any classroom is not likely to meet with success. Moreover, the increased diversity of undergraduate classrooms helps to increase the variety of viewpoints and experiences in a way that benefits all. University faculty facilitate learning for greater numbers of students if they also provide a diversity of experiences in their classrooms. As elementary and secondary education experiences become increasingly collaborative, students are primed for this type of interaction when they reach college.

SKILLS DEVELOPMENT

As is discussed throughout this report, graduates need a growing set of skills and competences to succeed in today's professional world. While content knowledge and technical skills will remain important, 21st-century students also need transferrable skills that will be useful in any career. Although agriculture graduates are not alone in needing these skills, the qualities that make agriculture colleges unique (see Chapter 1) make it especially appropriate that agriculture leads the way. These skills should be integrated throughout a curriculum and other student experiences rather than taught in separate courses. In fact, many of the strategies for teaching and learning discussed throughout this chapter can provide opportunities for developing these skills. Moreover, necessary experiences should be continual and extended over courses at all levels to allow additional achievement and growth throughout an undergraduate career. Departments and colleges are encouraged to conduct explicit planning to define how the skills will be incorporated into their academic offerings and how student achievement in these areas will be assessed.

Among the competences that students should develop are teamwork and working in diverse communities, working across disciplines, communication, critical thinking and analysis, ethical decision-making, and leadership and management. Those qualities are discussed briefly below.

Teamwork and Working in Diverse Communities

It is increasingly recognized that the challenges of the future will require the participation of many people working together in common pursuits. Yet college students rarely have the opportunity to engage in team-based activities as part of their academic work. Although many institutions offer collaborative activities (such as lab partners), they are generally intended more to extend resources than to afford educational experience. The committee believes that students should be provided with opportunities to work together both in and outside the classroom, to interact with and depend on people with different backgrounds, and to work on projects that will lead to better results than any student could have obtained alone. It will be especially important for students to gain experience in working with those who bring different backgrounds, skills, and perspectives. The workplaces of the future will be far more diverse than those of the past; students should be encouraged to gain multicultural awareness and to be comfortable in working with people of varied ages, ethnicities, and nationalities and with varied work styles.

Working across Disciplines

Closely related to working with others is the ability to work and speak across traditional disciplinary boundaries. Employers need their personnel not only to interact within disciplines but to bring expertise from different fields together to solve problems of common interest. As detailed in reports about interdisciplinary research and training (e.g., COSEPUP 2004), individuals in different disciplines often have trouble even speaking the same language. Agriculture colleges can and should help to prepare their students to speak not only to experts in their own field but more broadly with those in other fields and with the general public.

Communication

College graduates need excellent written and oral communication skills to work together, to speak to diverse audiences, and to communicate their knowledge and expertise more widely. Universities should provide all students with numerous opportunities to write and speak about a variety of topics to audiences that extend beyond their classmates. Students should be able to speak to those from other fields, from other countries, and from other sectors. Students should receive guidance and instruction by appropriate experts so that their communication skills improve. Peer review is especially encouraged: students should develop the skills and comfort not only to read and write their own work but to observe and critique that of others.

Critical Thinking and Analysis

Employers need workers who can make good decisions even when relying on data that are incomplete or even contradictory. Few academic institutions provide explicit training in critical thinking and analysis, and few classroom experiences challenge students in this regard. Moreover, mathematical analysis is often not incorporated into classes beyond a very basic level, and students have few opportunities to engage in quantitative reasoning. For example, students are rarely presented with real data or asked to suggest a strategy when the data do not point to a single "correct" answer. Textbook examples often downplay confounding data and simplify scenarios. Even laboratory and field experiences may involve some means of data cleaning so that students will be able to draw the "correct" inferences. The natural environment can make pedagogical activities more difficult, but it is vital that students have the opportunity to engage with real-world systems and to be forced to evaluate disparate data; they should be asked to make decisions on the basis of these data and to explain and defend their choices.

Ethical Decision-Making

Closely related to critical thinking and analysis is the need for students to make ethical decisions. That includes weighing sometimes contradictory aspects of disparate data and balancing competing interests. Professionals in all fields are asked to make tradeoffs all the time, and students need opportunities to hone their decision-making skills when they have appropriate guidance and the decisions are less critical. That type of thinking can easily be incorporated into classroom activities and assignments. For example, students could be asked to assess the risks and benefits associated with various practices to balance concerns coming from scientific, economic, environmental, and other arenas.

Leadership, Management, and Business

Skills that complement working in teams are motivating others and managing complex tasks, teams, and budgets. Students are almost never provided with formal opportunities to develop leadership and management skills. Many students assume leadership roles in extracurricular activities, but they are rarely given guidance on how to be an effective leader or manager or how to develop and work with budgets. Those skills are essential for surviving and thriving in the professional world, and the committee encourages institutions to build opportunities for students to hone them as part of their formal and informal undergraduate preparation. Closely coupled are facilitation and conflict resolution skills, which will enable teams and groups to work together effectively and to respond to challenges as they arise. Institutions should also look for opportunities to instill basic business and financial skills in their students.

CASE STUDIES AND PROBLEM-BASED LEARNING

Food and agriculture provide numerous real-world examples that can be brought into the classroom and used to enhance student learning, providing opportunities for students to practice the variety of transferrable skills described above. Case studies and problem-based learning provide ideal opportunities for students to work together in diverse teams, to consolidate information from a variety of disciplines, to communicate in both oral and written forms, to analyze data and evaluate evidence, and demonstrate leadership skills.

Problems taken from, or at least based on, actual experiences provide context and relevance to students (Capon and Kuhn 2004; Gijbels et al. 2005). Faculty can use examples from their own research, industry contacts and community organizations can propose challenges for courses, and extension activities can suggest issues of concern in a given state (see Chapter 5 for a discussion of the role of outreach, extension, and industry connection in fostering undergraduate education).

Those types of cases and problems also provide opportunities for students to learn by doing. They may even be able to contribute to solutions to the real-world problems that they are given. For example, one of the posters presented at the summit described a capstone experience at California Polytechnic State University that has students working on real problems of commercial interest (Box 3-3). Case studies and problem-based learning can help students to understand why academic knowledge matters.

BOX 3-3
Learning by Doing at California Polytechnic State University

Many agriculture colleges stress a "learn by doing" pedagogy designed to better prepare students and allow them to demonstrate competency. California Polytechnic State University in San Luis Obispo has had experiential learning as a trademark for over 100 years, with ample opportunities for students to incorporate internships, laboratory classes, and capstone senior thesis projects into their curriculum. An "Enterprise Project" option gives students access to learning in the context of commercial projects in livestock, fruit, vegetable, and honey production. After they have completed coursework on the topic, students are given responsibility in one of the commercial enterprise project areas under the supervision of a faculty member. The goal of the project is to be profitable and for students to gain credit, practical experience, and potentially a share of the profit. Additional benefits are a stronger work ethic, sense of accomplishment, experience in teamwork, analysis, synthesis, and assessment.

Poster presented at Summit: Jonathon L. Beckett, Lynn E. Moody, Mary A. Whiteford, and Mary E. Pedersen. "Learn by Doing Pedagogy in Agriculture through Enterprise Projects."

SERVICE LEARNING AND COMMUNITY ENGAGEMENT

Agriculture lends itself to what has been termed "service learning," in which students learn and receive academic credit for participation in activities that meet community needs (Astin et al. 2000; Battistoni 2001; Gelmon et al. 2001).[2] One could see service learning as the intersection of community service and academic study. By drawing on scholarship in the natural and social sciences, civic engagement helps to make content knowledge come alive and allows students to contribute to the needs of their community. Service learning also helps to connect the university with the community. Testifying to the importance of recent developments in academic–community interactions, the Carnegie Foundation for the Advancement of Teaching has established an elective classification in community engagement:

> **Community Engagement** describes the collaboration between institutions of higher education and their larger communities (local, regional/state, national, global) for the mutually beneficial exchange of knowledge and resources in a context of partnership and reciprocity.[3]

Particular elements of the Carnegie classification include the engagement of faculty, students, and community in mutually beneficial and respectful collaboration that addresses community needs, deepens student learning, enriches scholarship, and enhances community well-being. Such activities can focus on outreach—applying institutional resources for community use—or partnership—in which collaborative interactions are common. These are some of the same elements that the committee highlights as integral to successful partnerships in Chapter 5.

Community engagement and service learning are natural outgrowths of many of the best practices discussed throughout this report. Because agriculture encompasses many areas of study and application with obvious community connections, the committee hopes that agriculture colleges will take advantage of opportunities for students to engage with their communities and receive academic credit for service learning. Several institutions have taken substantial steps to incorporate service learning throughout their campuses (see Box 3-4 for one example).

---

[2]Campus Compact serves as a clearinghouse for engaging students in service learning. See <http://www.compact.org/> for more information.

[3]See <http://www.carnegiefoundation.org/classifications/index.asp?key=1213> for more information.

BOX 3-4
Center for Excellence in Curricular Engagement at North Carolina State University

North Carolina State University (NCSU) established a center devoted to service learning and curricular engagement in 2007, integrating the institution's land-grant mission with a commitment to educational innovation and leadership development. The NCSU Center for Excellence in Curricular Engagement hopes to expand community-engaged teaching, learning, and scholarship at the university; collaborate with other institutions to advance curricular engagement throughout North Carolina; and establish NCSU as a leader in curricular engagement. The center offers consultation and development opportunities for faculty and workshops for all members of the university community, promotes the scholarship of teaching and learning on campus and beyond, and partners with campus- and community-based organizations to enhance and create opportunities for community-engaged learning.

Additional information about the center is available at <http://www.ncsu.edu/curricular_engagement/>.

COOPERATIVE AND ACTIVE LEARNING

Cooperative learning began in elementary schools in the late 1960s largely through the research and efforts of Robert Slavin, Elizabeth Cohen, Spencer Kagan, and David and Roger Johnson. As a result of recognition that cooperative learning is an effective teaching and learning strategy for higher education, it appeared on the college instruction scene in the 1990s. Cooperative learning is more than just group work; it incorporates several elements: positive interdependence, face-to-face interaction, individual accountability, interpersonal skills, and group processing (Johnson and Johnson 1989; Johnson et al. 1991; McNeill and Payne 1996; McNeal and D'Avanzo 1997; Michaelsen et al. 2002). Cooperative learning often involves specially prepared lessons in which well-formed groups approach questions that are designed for teamwork.

In one example of this type of cooperative learning, Beichner et al. (1999) have pioneered the SCALE-UP project, in which classes of up to 100 students are taught by dividing the students into small groups whose members work collaboratively with each other and with other groups in classrooms redesigned for collaborative work.[4] Such studio classrooms make

[4]See <http://scaleup.ncsu.edu/> for more information about SCALE-UP.

it easier for students to work together, turning the classroom from an instructor-centered to a student-centered environment. Classroom architecture that supports—rather than impedes—cooperative learning would help to break down some barriers to active learning. Although the committee does not expect universities suddenly to dedicate millions of dollars for classroom renovations, it hopes that universities will seriously consider pedagogy and instructional needs as part of the planning for new construction and renovation. That is, if an institution is building or renovating a building that includes classrooms, the committee hopes that the instructional spaces will be aligned with learning objectives and with the types of instruction that could be incorporated into the spaces.[5]

Collaborative and active learning also includes a variety of less formalized arrangements, including tasks on which groups of students work together over only a minute or two. For example, the Peer Instruction technique developed by Eric Mazur (1997) intersperses a traditional lecture class with a series of ConcepTests, short conceptual questions that students consider individually before discussing them in small ad hoc groups and trying to convince each other of their answers. Results show that students not only answer correctly immediately on reconsideration of a question after small-group discussion but retain the knowledge until the end of the term (Crouch and Mazur 2001).

Numerous research studies and meta-analyses show that students learn more from teaching methods in which they are actively engaged than from traditional lecture formats (Hake 1998; Wright et al. 1998; Fagen et al. 2002; Knight and Wood 2005; Michael 2006; Armstrong et al. 2007). Therefore, it might seem surprising that so much of science instruction, including instruction in agriculture, depends on passive lecture courses. As active learning slowly becomes more common, the role of faculty changes; as students take more responsibility for their own learning, faculty may serve more as guides and facilitators than as the providers of knowledge (Kirschner et al. 2006).

LEARNING COMMUNITIES

In recent years, several colleges and universities have established learning communities that bring together students through connected coursework

---

[5]Project Kaleidoscope has been advising institutions on developing facilities that support teaching and learning. Resources are available at <http://www.pkal.org/collections/PKALFacilitiesResource.cfm>.

around a theme or major, often including themed residence halls. The general purpose of these learning communities is to create a sense of camaraderie and shared experience, especially in environments in which the size of the university may intimidate (Shapiro and Levine 1999; Taylor et al. 2003; Laufgraben and Shapiro 2004; Smith et al. 2004). The settings also provide teaching and learning venues outside the normal classroom, allowing a variety of instructional strategies that can address different learning styles and provide multiple assessment opportunities that facilitate learning.

Learning communities in which there is a focus on the agricultural sciences have been developed in many universities and colleges. Models that are residential, academic, or a combination of the two exist at such universities as Auburn, Colorado State, Iowa State, Minnesota, Missouri, Nebraska–Lincoln, New Mexico State, Ohio State, Oklahoma State, Purdue, Tennessee, Texas A&M, and Texas Tech. For example, students at Purdue have several agricultural learning-community choices in which to enroll: Agricultural Education; Animalia; and Wood, Water and Wild Wonders. The benefits are varied and are both academic and nonacademic; for example, the retention rate is almost 5% higher among students who participate in learning communities at Purdue than among other students.[6]

EXTRACURRICULAR ACTIVITIES

The undergraduate experience consists not only of coursework and associated formal academic responsibilities but of extracurricular activities and other aspects of college life. For many students, the structured curriculum makes up only a small part of what they find valuable in a college experience. In fact, these noncourse experiences are likely to be more influential than formal course experiences in their career decision-making. Student organizations and the sense of responsibility that often comes with them can be important motivators and influences for students and should not be undervalued. Students may also have extracurricular opportunities to connect directly with issues related to food and agriculture, such as supporting school farms or community gardens, working with an institution's dining service to influence menus and purchasing, and volunteering in the community to address issues of nutrition, hunger, and obesity.

---

[6]See <http://www.purdue.edu/sats/learning_communities/instructors/facts/success.html> for more information.

UNDERGRADUATE RESEARCH

Research experiences for undergraduates (REU) can provide students with the opportunity to contribute to original research, to gain first-hand experience in conducting research, and to participate in laboratory communities. These experiences have been helpful in retaining students in their disciplines so that they complete science degrees and pursue graduate study at a higher rate (Bauer and Bennet 2003; Kardash 2000; Lopatto 2003, 2004, 2007; Rueckert 2002; Hunter et al. 2007; Russell et al. 2007). Such experiences can take many forms, including independent studies, senior theses, research or laboratory courses, substantive labs supplementing existing courses, short-term experiences during vacations or January terms, or even having students holding part-time jobs supporting faculty laboratories. As discussed in Chapter 6, the committee hopes that it will become common for undergraduate students in food and agriculture programs to have the opportunity to participate in research; achieving this aim will require support and facilitation by universities and funding agencies as such experiences often require significant personnel and financial resources.

REU programs have been quite common in the basic sciences; the National Science Foundation (NSF), the National Institutes of Health, and the Howard Hughes Medical Institute (HHMI), for example, support such initiatives. Some REU students participate in formal institutional summer programs that bring in a cohort of students to conduct research with a variety of faculty. Other REU opportunities are made available by individual laboratories, often supported by supplements to existing research awards funded by an agency. Agriculture-focused funding agencies, such as the U.S. Department of Agriculture, can learn from the experiences of the other agencies in developing funding opportunities and providing REU supplements to facilitate undergraduate research opportunities. Of course, agriculture students can also be encouraged to participate in *existing* REU programs—including those outside of agriculture—to gain experience in the field or laboratory and an appreciation for what research is.

INTERNATIONAL EXPERIENCES AND PERSPECTIVES

The increasingly international nature of agriculture suggests the need for students to have greater exposure to international perspectives. Such opportunities can take the form of learning-abroad programs—in which students spend a semester or more studying in another country—and by increasing the international content in courses at U.S. institutions. At present, however, participation in such experiences is relatively rare.

Learning-Abroad Programs

Most institutions offer some type of "study-abroad" program for their students; for example, the American Council on Education (ACE) estimates that 91% of institutions offer opportunities in education abroad (Green et al. 2008). The percentage of students participating is far lower: the same ACE survey found that more than one-quarter of institutions had no students graduating in 2005 who had studied abroad.

Learning-abroad programs are excellent in improving cultural sensitivities and increasing understanding of another country and its language, and the committee fully supports these goals. Yet there also is an opportunity to supplement these general programs with targeted opportunities for interested students that combine general cultural immersion with experiences focused on the global agricultural infrastructure. Even many institutions with a strong tradition in food and agriculture have not featured these fields in their international programs.

In recent years, however, several programs have started to address agricultural topics with a multidisciplinary approach focused on a specific commodity or a specific region (see, for example, a description of the Michigan State University programs in Box 3-5). That approach provides a contextual view of issues while using a manageable framework as its basis. Such programs increase the depth of students' understanding and prepare them for future roles, whether in academe, government, industry, or other sectors.

Of particular utility to institutions looking to expand their international programs may be the recently announced Center for Capacity Building in Study Abroad, a joint project of NAFSA: Association of International Educators and the Association of Public and Land-grant Universities.[7] The center, launched in 2008, supports learning abroad by identifying opportunities in emerging and high-demand study-abroad markets, helping institutions to access these markets, building a database to support institutional expansion efforts, and fostering information-sharing among institutions.

[7]See <http://www.studyabroadcenter.org/> for more information about the Center for Capacity Building in Study Abroad.

BOX 3-5
Michigan State University International Programs

Michigan State University has one of the country's most extensive international-program offerings, including 50 study-abroad programs in the College of Agriculture and Natural Resources (CANR). The university as a whole has made a strong commitment to internationalization, even featuring those programs as a central element of its 2006 reaccreditation self-study. As explained in a presentation at the Leadership Summit, the university recognized that internationalization means not only engaging in outreach to other counties but encouraging true discussion and understanding of different ways of doing things and of different belief systems.

CANR's study-abroad opportunities have great variety and are often focused on particular regions of the world and their local concerns. For example, the Conservation and Biodiversity in Parks and Nature Reserves in South Africa summer program provides students with perspectives on land management by considering the effects of land-based activities and international policies on the natural communities in these ecosystems, including the role of game reserves, nature reserves, and national parks as management tools. The program addresses both scientific and social issues, such as what happens when restrictions to protect biodiversity are imposed on a society and the effect of hunting in private game reserves on the surrounding communities.

Another example is a community-engagement program in rural Ireland, in which students not only have an immersive living experience but work with local leaders to foster community development activities in the Tochar Valley. The program provides students with real-world, practical experience and direct connection to people working in their own communities.

Additional information about CANR's international programs is available at <http://www.canr.msu.edu/overseas/>; see <http://www.accreditation2006.msu.edu/internationalization/index.html> for information about the Internationalization Self-Study.

Summit presentation: Frank Fear, *Senior Associate Dean*, and Paul Roberts, *Director, Study Abroad and International Training, College of Agriculture and Natural Resources, Michigan State University.*

International Perspectives in U.S. Course Content

Increasing the inclusion of internationally based lecture topics, case studies, and research programs in existing structures would deepen students' understanding of international perspectives and the increasingly interconnected food and fiber supply chain. Despite the benefits, few institutions require students to take international-focused courses; ACE reports that only

BOX 3-6
Globalization of the Science Classroom at the University of Maryland

A number of courses at the University of Maryland, College Park, are working to add a global perspective with support from the Freeman Foundation. The East Asia Science and Technology (EAST) program seeks to introduce East Asian themes into a variety of undergraduate science and engineering courses, including honors seminars, required introductory courses, and general-education offerings. Faculty participants are named as EAST fellows and provided with support to develop new courses, work with colleagues abroad to develop global courses, and create modules to introduce an East Asian perspective into existing courses.

EAST program courses incorporate a number of interactive pedagogical elements—such as active learning, problem-based research, team-based learning, and student peer review—and opportunities to engage with experts on East Asian issues. At the time of the Summit, 13 EAST fellows were offering 18 courses that reached some 1,600 students. Two of the courses were transnational and were offered concurrently at the University of Maryland and an East Asian university.

In addition to the courses, the EAST program incorporates exchange of students and faculty and numerous collaborations in curriculum design and research.

Summit presentation: Robert Yuan, *Professor Emeritus of Cell Biology and Molecular Genetics, University of Maryland, College Park*; Vanessa Sitler, *senior undergraduate student in business management, Robert H. Smith School of Business, University of Maryland, College Park.*

37% of institutions require a course with an international or global focus, according to a 2006 survey (Green et al. 2008).

The committee believes that students should have the opportunity to be exposed to global perspectives even without leaving their U.S.-based classroom. Forming unique partnerships with foreign universities might be one way to encourage instructors and students to engage in those topics and potentially to collaborate on issues of mutual interest and identify new approaches to searching for solutions (see Box 3-6 for an example), but more modest ways to incorporate international perspectives will also have value.

INSTRUCTIONAL TECHNOLOGY

Advances in technology are helping to move education from hour-long class meetings several times a week to round-the-clock continuous "virtual learning communities." They also help students become comfortable with using a variety of technologies that may serve them well in future endeavors. Minimal uses of technologies—such as static course Web sites and e-mail contact with instructional staff—have been around for more than a decade, but innovative forms of *instructional* technology can transform the educational environment and lead to substantial changes in how information is transferred, how students interact with their teachers and fellow students, and even how students "attend" class.

For example, expansive course Web sites enable students to interact with each other and with instructional staff, extending classroom interactions to 24 hours a day (e.g., Colbert et al. 2007). Commercials tools such as Blackboard make Web sites easier for even the most technophobic faculty member to develop and maintain. High-speed Internet connections and advances in videoconferencing technology make it possible to link classrooms from around the world virtually and provide real-time international perspectives without requiring any travel.

To be sure, some instructional technology merely provides a mechanism for conducting activities more efficiently than but not fundamentally differently from nonelectronic approaches. For example, providing access to course material and lecture notes or even allowing homework submission on course Web sites may be valuable but does not break new pedagogical ground. However, technology can enhance the ability of instructors to conduct formative assessment, such as in Just-in-Time Teaching, in which students provide feedback to their instructors a few hours before class by answering questions posted online (Novak et al. 1999).[8] Other uses of technology enhance the learning process in ways that were not previously possible. Students can now perform simulations, collect and analyze data, and tap into information collected by others whenever they have access to a computer; this enables them to engage more directly with original scholarship and experimentation without expensive laboratories or the challenges introduced by large classes. Moreover, in today's connected culture students can do all those things while sitting in the classroom, in the coffee shop, in their dormitory, in their pajamas, and in places around the world.

Educational technologies also enhance students' in-class experiences. For example, "clickers" allow students to respond to questions posed by an instructor and provide feedback to instructors on student understanding in real time (Beatty et al. 2006; Fies and Marshall 2006; Barber and Njus 2007; Caldwell 2007; Preszler et al. 2006; Bruff 2009). These wireless devices

[8]See <http://www.jitt.org/> for more information about Just-in-Time Teaching.

provide a mechanism for instructors to get instant and anonymous feedback from students during class. Typically, an instructor will pose a carefully designed multiple choice question to the class, and students will respond with their best answer to the question. If students overwhelmingly answer correctly, the instructor can move on the next topic, with the reassurance that students are on board; but if student responses reveal lack of understanding, the instructor has the opportunity spend more time discussing the confounding topic right then. This formative assessment in the classroom allows faculty to be in more nearly constant touch with what students are learning, not only what they are teaching—and to do so in real time, not weeks later on the midterm examination. In the words of University of Colorado biologist Bill Wood, clickers may have become the "greatest new teaching tool since chalk."

Distance learning is growing as an industry and as a way for students to obtain educational experiences on their own schedules and without leaving their homes. For example, nearly 20% of those enrolled in degree-granting postsecondary institutions take at least one online course (Allen and Seaman 2007). It is likely that land-grant universities will be especially called on to expand and enhance their online offerings in an effort to serve the populations of their states more efficiently and to enable course enrollment without requiring student to take courses or even set foot on campus. Distance education is even being used to support extension activities. Early in 2008, for example, a consortium of 74 land-grant universities launched eXtension, a national Web site that provides farmers not only resource information—similar to what they received through state extension networks—but opportunities for collaboration and communication on a wide variety of issues (Guess 2008).[9]

IMPLEMENTING CHANGE

Developing an undergraduate experience that integrates the skills and experiences discussed throughout this report will require the attention of a wide variety of stakeholders. Resources will also be required for some interventions, but it is important that institutions not use a lack of resources as an excuse not to act.

In some cases, funding agencies may be in a position to implement new programs to target specific educational innovations. In other cases, agencies may be able to provide relatively small supplements to existing grants

[9]See <http://www.eXtension.org/> for more information about eXtension.

to achieve those aims. For example, the National Science Foundation (NSF) sponsors many undergraduate research experiences by providing supplements to NSF-funded research awards made to individual investigators (the agency also provides funding for dedicated programs that offer research experiences for undergraduates). Agencies can encourage the development of educational activities that leverage the support already provided to researchers. To that end, one of the criteria used by NSF in making awards is the "broader-impacts criterion," which includes contributions to teaching and learning, broadening the participation of underrepresented groups, enhancing the infrastructure for research and education, disseminating results broadly, and providing societal benefits.[10] Those approaches may serve as models for other agencies and private sponsors to think about ways to encourage the development of best practices in teaching and learning with relatively modest investments. When institutional grants and other funds are available, the committee hopes that deans, department chairs, and other administrators take advantage of the opportunity to support and encourage such goals as high-quality teaching, active and service learning, extension and outreach, and international experiences.

Although new and external funding will certainly help institutions in effecting change, the committee strongly argues that institutions need to take the necessary steps even if additional funding is not available. Institutional priorities will need to emphasize undergraduate education, and universities may need to make tough decisions about redirecting support from other programs.

ADOPTION OF EFFECTIVE TEACHING METHODS

Despite decades of research demonstrating the effectiveness of teaching methods, including active student engagement, adoption by individual faculty has been slow. That suggests that one of the most important challenge in reforming teaching and learning is not basic knowledge of what works but putting the information in the hands of faculty, providing the necessary infrastructure, and providing the appropriate incentives for faculty to implement the methods.

Most faculty are not aware of the research on teaching and learning, because it is not a formal part of most graduate training. They enter profes-

[10]See <http://www.nsf.gov/pubs/gpg/broaderimpacts.pdf> for examples of activities that are responsive to NSF's broader-impacts criterion.

sorships without much pedagogical knowledge and often revert to teaching how they have been taught, which often means that undergraduate classes—and especially those at the introductory level—tend to be lecture-based passive environments. As will be discussed below, faculty development can provide a mechanism for enhancing knowledge of research in teaching and learning.

Although it is important, simply telling faculty about education research is unlikely to be sufficient to effect change. Lack of information is only one of the barriers to the implementation of research-based teaching methods. For example, Henderson and Dancy (2007) found a number of situational barriers that limit education reform, including student attitudes and preparation, expectations of content coverage, limited instructor time, departmental norms, student resistance, class size, classroom layout, and structure of instructional time. Those barriers present an important challenge that will need to be addressed. Even bringing the challenges into the open can have a powerful effect in encouraging faculty to overcome them. As one mechanism, centers for teaching and learning could offer faculty development opportunities and discussions in which faculty can work together. Providing opportunities for science faculty to interact and work more closely with education researchers also appears to help in implementation (Henderson and Dancy 2008).

ROLE OF GRADUATE EDUCATION

This report is focused on undergraduate education, but graduate students play an important role as well. Graduate students serve as teaching assistants and often have more contact with undergraduates than do members of the faculty. They serve as mentors in research and teaching laboratories. They have viewpoints and experiences that can be helpful in curriculum development and are often less confined to a single discipline or field of study than are faculty. Perhaps most important, graduate students—and postdoctoral researchers—are the faculty of the future. Therefore, engaging them in conversation about the reform of undergraduate education while they are still students and trainees will pay off for years to come. And it will make those graduate students and postdocs more valuable on the job market if they can demonstrate depth in their thinking about teaching and learning (see Box 3-7 for an example of a program designed to help graduate students to be effective teachers).

BOX 3-7
Enhancing Graduate Training in Teaching and Learning: Delta Program at the University of Wisconsin

The Center for the Integration of Research, Teaching, and Learning (CIRTL) at the University of Wisconsin–Madison has a goal of developing a national faculty in the natural and social sciences, engineering, and mathematics with the knowledge and experience to forge successful professional careers that include implementing and advancing effective teaching and learning practices.

Building on a prototype at UW–Madison, CIRTL's Delta Program now connects six research universities in a curriculum of graduate courses, intergenerational small-group programs, and internships embedded within an interdisciplinary learning community. Every facet of Delta is designed around models familiar to researchers in these disciplines. For example, the courses are project-based, and require students to define a learning problem; understand their student audience; explore the literature for prior knowledge; hypothesize, design, and implement a solution; and acquire and analyze data to measure learning outcomes. Delta internships are research assistantships in teaching, in which a graduate student or postdoctoral researcher partners with a faculty member to address a learning problem. Delta activities are also designed to provide each graduate and postdoctoral participant with a portfolio, letters of recommendation, and presentations/publications in teaching and learning analogous to those in their disciplinary research *curriculum vitae*. Since 2003, more than 1,600 UW–Madison graduate students, postdoctoral researchers, staff, and faculty have participated in the Delta learning community.

The Delta Program has enabled graduate students and others early in their careers to develop the skills and confidence they need to become creative, well-prepared professionals who will enter the national workforce with the ability to teach effectively and improve science education broadly.

Additional information about CIRTL is available at <http://www.cirtl.net/>; information about the Delta Program is available at <http://www.delta.wisc.edu/>.

CENTERS FOR TEACHING AND LEARNING

A number of institutions have established centers dedicated to improving undergraduate instruction. Whether they are called centers for teaching excellence, centers for teaching and learning, or something else, they are typically staffed by education professionals who work with faculty, graduate teaching assistants, and others to improve undergraduate education (Singer 2002).

Implementing several of the ideas discussed in this report might be best carried out by such centers. They already provide an existing infrastructure in a local setting, have resources and expertise to conduct workshops and other activities, and tend to have a campuswide reach. In addition, they can serve as a valuable resource to an institution by providing individual consultation and programming to those seeking to improve teaching on their campus.

FACULTY DEVELOPMENT

Implementing the changes that would promote effective teaching and learning in undergraduate agricultural education and support the success of a diverse student population will require adequate resources. Although support for academic research is often available to faculty from external and internal sources, few resources are available for teaching. Educational innovation may be relatively inexpensive, but it is not without some costs: faculty need resources to enhance their teaching, to develop new courses, or to learn new teaching techniques. Teaching assistants (TAs) can often be critical in enabling faculty to implement new teaching techniques, and their support can often be provided through instructional budgets. TAs not only help to take on time-consuming responsibilities, but involving TAs in educational innovation offers an excellent opportunity to provide graduate students and others with professional development experiences. Resources are also needed to allow faculty to keep up with the scholarship and practice of undergraduate education: support for books, journal subscriptions, society memberships, and participation in relevant meetings and conferences. Perhaps the scarcest resource for many faculty is time itself, and release time may be an appropriate investment in curriculum reform.

Faculty development, in general, is essential for helping to prepare faculty to take advantage of the research on teaching and learning. Faculty development will have to be multifaceted to include both formal training and support for participation in ongoing networks; it should occur at several levels and be conducted by a variety of communities. Universities have an obvious responsibility to ensure that their faculty are kept current with research on teaching and learning, new pedagogical techniques, and developments in instructional technology. Professional societies have an important role to play in supporting high-quality education in their disciplines and can bring expertise in teaching and learning to individual fields of study. Funding agencies, accrediting bodies, and other national organizations can help to promote and support activities to convene faculty to discuss these issues and promote the scholarship of teaching and learning. Relatively inexpensive investments in such activities will pay dividends for years to come.

There are many models of faculty development, from individual workshops to year-long sabbaticals focused on teaching and learning. It is likely that in-depth—but brief—experiences provide maximal benefit for a small investment of time or resources. For example, the National Academies has developed a week-long faculty-development institute for undergraduate faculty in the biological sciences with a particular focus on research universities (see Box 3-8). Another type of model is a network of faculty dedicated to a common purpose, such as Project Kaleidoscope's Faculty for the 21st Century network (see Box 3-9).

BOX 3-8
National Academies Summer Institute on Undergraduate Education in Biology

The authors of the 2003 National Research Council report *Bio2010: Transforming Undergraduate Education for Future Research Biologists* recognized the central role of faculty development in effecting changes in undergraduate education, and they devoted one of their eight recommendations to campus-level and national faculty development (NRC 2003a).

As a direct result of that recommendation, the National Academies established the National Academies Summer Institute on Undergraduate Education with support from HHMI, the Research Corporation for Science Advancement, the Presidents' Committee of the National Research Council, and the University of Wisconsin–Madison (Wood and Gentile 2003; Wood and Handelsman 2004; Pfund et al. 2009). The summer institute seeks to transform undergraduate biology education at research universities nationwide by improving classroom teaching and attracting diverse students to science. Teams of two or three faculty members, most of whom teach introductory courses, learn about and implement the themes of "scientific teaching" (Handelsman et al. 2004)—active learning, assessment, and diversity—during a week-long workshop dedicated to teaching and learning. Participants work together to develop materials and lessons that they agree to implement in their courses in the following year.

The impact of the summer institute is far greater than the individual teaching materials; rather, it seeks to transform how individual faculty members view their teaching and, by extension, influence other members of their departments and their disciplines to make similar transformations (Pfund et al. 2009). Participants are named National Academies Education Fellows in the Life Sciences and are encouraged to become ambassadors for education reform on their campuses and throughout their professional communities. The aim is, therefore, to leverage a program that directly reaches 40 faculty per year—who themselves teach over 15,000 students per year—into one that reaches hundreds of thousands of students.

Additional information about the Summer Institute is available at <http://www.AcademiesSummerInstitute.org>.

BOX 3-9
Project Kaleidoscope Faculty for the 21st Century

Since 1994, Project Kaleidoscope (PKAL) has been managing a national network of emerging leaders in undergraduate science, technology, engineering, and mathematics (STEM), known as Faculty for the 21st Century (F21). The network encourages faculty to become agents of change and visible leaders on their campuses and in their disciplines.

The goal of the PKAL F21 network is to foster every F21 member's capacity for leadership by providing opportunities to explore new ways of thinking about students, about science and technology, and about society. PKAL intends to build a supportive alliance among and between the F21 members and the affiliated network of current leaders in STEM education. The F21 network now includes more than 1,200 faculty at over 500 colleges and universities around the country.

F21 members are nominated by senior administrators on their campuses, who must make a commitment to enhance the leadership capacities of their nominees. The collaboration between PKAL and participating campuses is an essential ingredient of the F21 network in recognition that groups working together can accomplish more than those working in isolation.

Additional information about the PKAL F21 network is available at <http://www.pkal.org/activities/F21.cfm>.

Summit Presentation: Jeanne Narum, *Director, Project Kaleidoscope.*

The committee believes that institutions should include teaching-focused workshops and experiences as part of graduate education and postdoctoral training. Graduate students and postdoctoral scholars make up the next generation of faculty, and early intervention in their training can lead to faculty who are already familiar with education research and comfortable with student-centered pedagogies when they begin their faculty careers. One national effort that strives to prepare graduate students for careers at a variety of academic institutions with a variety of missions, student bodies, and faculty expectations is the Preparing Future Faculty (PFF) initiative.[11] Individual PFF programs address the full scope of faculty responsibilities; provide multiple mentors to students, including mentors in teaching; and engage a cluster of diverse institutions so that students have opportunities to work with faculty and gain teaching experience in a variety of settings.

[11]See <http://www.preparing-faculty.org/> for more information about the PFF program.

FACULTY REWARDS

One of the greatest obstacles to the reform of teaching and learning cited at the Leadership Summit was the institutional reward structure, especially the criteria for promotion and tenure. A thorough review of institutional tenure-review policies is far beyond the scope of this report, but the committee believes that the importance of the issue merits a brief discussion here.[12]

There was a strong feeling among participants in the summit that tenure criteria are strongly tilted toward faculty members' research productivity and that too little attention is paid to teaching and service. Faculty, understandably, are driven by what their employers value: in the current reward structure, this means research activities, especially being published and securing external grant support. Even though teaching and learning are at the heart of academic institutions, they rarely play a substantial role in the evaluation of faculty. In part, that is because of the perceived difficulty in measuring teaching quality objectively, but there are strategies for evaluating faculty teaching and student learning (NRC 2003b). Many institutions do offer some sort of teaching award, but some complain that such awards can be little more than popularity contests that reward entertaining or dynamic instructors. The competitions are often based solely on student evaluations and rarely consider measures of student learning. Moreover, an institution may give only a handful of the awards each year, leaving many excellent instructors without recognition or acknowledgment.

Implementing high-quality educational practices and enhancing institutional rewards for teaching and learning will require renewed emphasis at all levels, including the top of an institution. When the driving force for the process flows from the president and provost, the attention of internal and external stakeholders can be focused on support and encouragement for teaching. Some of the changes that will be required are a refocusing of faculty hiring and evaluation to include consideration of learning outcomes, valuing the scholarship of teaching and learning in the promotion and tenure process, and adopting other strategies for honoring and supporting teaching.

A number of ideas for rewarding undergraduate teaching and supporting student learning were offered at the Leadership Summit. Some institutions have created teaching tracks in which instructors are judged primarily on the basis of the quality of their teaching and that are separate from the

---

[12]For a more thorough discussion of institutional rewards for teaching, see, for example, NRC (2003b).

research-track faculty that have been the standard. The positions sometimes have distinct titles, such as "professor of the practice of . . . ." In most of the cases discussed, however, the teaching track does not offer the possibility of tenure, and there are often limitations on involvement in faculty governance; this raised concerns about a two-tier system in which teaching faculty are relegated to a lower rank. Some institutions have established tenure-track faculty positions in discipline-based education, affording faculty the same opportunities and responsibilities as traditional research-focused faculty members. Those holding such positions are expected to conduct original research, publish in peer-reviewed publications, secure extramural funding, and become leaders in their fields; the only difference is that their research is focused on education.

One institution participating in the summit had come to the conclusion that teaching should not be considered the *individual* responsibility of faculty members but the *collective* responsibility of an entire department. That change in mindset helped to encourage an open discussion of teaching and learning at that institution instead of something that happened behind closed doors. The institution even decided to offer teaching awards to entire departments; in addition to public recognition, the award comes with a prize of unrestricted funds that the department can spend as it sees fit. Because such unrestricted funds are so uncommon at most institutions, this can be a powerful motivator for even a recalcitrant department to focus on teaching quality and student learning.

Perhaps most interesting were institutions that had incorporated undergraduate teaching into their tenure criteria (see overview in Bush et al. 2006). One speaker described an extensive plan at the University of Wisconsin-Madison (UW–Madison) that pays attention to teaching in tenure consideration. As explained in Box 3-10, UW–Madison has developed a structure in which teaching can be a primary area of accomplishment for tenure consideration; it can also serve as a secondary area that is taken seriously for faculty who have research as their primary area of focus. Policies like those at UW–Madison can serve as models for other institutions in drafting similar criteria that give appropriate consideration to teaching. What may be more of a challenge, however, is getting universities to both adopt and enforce such policies.

Some concern has been raised about the danger of creating a "caste system" in which some faculty concentrate on teaching and others are committed mostly to research. Institutions that choose to pursue such a path will need to ensure that compensation, advancement, job security, and respect are provided equally to teachers and researchers.

BOX 3-10
Valuing Teaching for Tenure and Promotion at the University of Wisconsin–Madison

The University of Wisconsin–Madison has taken steps to value undergraduate teaching in the criteria for granting tenure. As described by Caitilyn Allen, professor of plant pathology, who served as chair of the university-level tenure committee in 2005–2006, UW–Madison made a commitment to incorporate a rigorous and fair evaluation of teaching for consideration of tenure. The university has established a culture that has the support of the administration; this means that department decisions based on teaching cannot be outweighed by concerns about external grant support.

Tenure dossiers at UW–Madison must describe achievements in research, teaching, outreach (extension), and service; a candidate must show "excellence" in one and "significant accomplishment" in a second; any candidate with a teaching appointment is judged partly on the basis of teaching. Those seeking tenure primarily on the basis of teaching must demonstrate a national or international reputation that is demonstrated by scholarly work related to teaching. More commonly, teaching is considered an important accomplishment to support a primary focus on research or extension.

Among the metrics used to evaluate teaching are the following:

- Numbers of courses and students taught, taking into consideration how many were new preparations and student mentoring outside class.
- Student evaluations, including numerical ratings for each course, but also qualitative student evaluations for a selection of courses and exit interviews with a handful of randomly selected students from each course. Even if individual student responses are not always objective or fair, the collective wisdom of many students usually provides an accurate picture of an instructor.
- Peer review, in which two faculty members observe two class sessions each semester and write an evaluation that is discussed with the junior faculty member. For those seeking tenure primarily on the basis of teaching, an independent committee of master teachers from outside the home department is brought in to assess the candidate's teaching.
- Evaluation of teaching materials, including a two-page statement of teaching philosophy and practice, new curriculum development (when relevant), and copies of original teaching materials, such as syllabi, assignments, examinations, and laboratory or field exercises.
- Measures of the effect of the candidate's teaching-related work beyond his or her own classroom, including peer-reviewed articles, textbooks, and other ped-

agogical materials; presentations at regional, national, and international meetings; grants to develop courses or curricula or to conduct pedagogical research; and documentation that the candidate's teaching activities have resulted in changed practices beyond the campus.

- Letters from off-campus experts on teaching in the candidate's field who review and assess the faculty member's teaching dossier in the case of someone being considered primarily on the basis of teaching.

In many ways, those measures are directly comparable with measures used for research productivity, so they should not be foreign to most faculty.

Allen noted that the reviews are not pro forma but are taken seriously. There are examples of candidates with good external funding and substantial publication records who failed to be promoted because of the absence of high-quality teaching.

The university coupled the altered procedures with institutional resources to support faculty as teachers: partnering young faculty with master teachers and developing mentor committees, providing peer review and comments on teaching, offering workshops and symposia to generate ideas and build a culture of teaching, granting teaching leaves to assist in course development and revision, and awarding small grants for computers and software, memberships, subscriptions to teaching-oriented publications, and attendance at education-focused conferences. And it has taken steps to continue to value teaching after tenure, for example, by making it one criterion for annual merit salary adjustments, requiring sabbatical applications to include a justification for teaching development, nominating instructors for teaching awards, and publicizing faculty teaching accomplishments.

For additional information see the University of Wisconsin–Madison's Guidelines for Recommendations for Promotion or Appointment to Tenure Rank in the Biological Sciences Division at <http://www.secfac.wisc.edu/divcomm/biological/Tenure-Guidelines.pdf>.

Summit Presentation: Caitilyn Allen, *Professor of Plant Pathology, University of Wisconsin–Madison.*

Support for teaching can also be incorporated into faculty hiring. Actions such as emphasizing and providing appropriate descriptions of teaching opportunities in position descriptions, asking for statements of teaching philosophy and experience as part of an application, discussing teaching and learning during interviews, asking candidates to conduct a sample class during a campus visit, and involving students in the campus visit can send a signal that teaching is valued and provide information that hiring committees can use in assessing a candidate's teaching ability. Institutions could also consider devoting a portion of startup costs to education-related expenses; even a small amount of money can go a long way in emphasizing the importance of teaching and providing the impetus for faculty to learn more about effective teaching strategies or teaching materials. Those steps, if taken early, can help to reinforce attention to education that can last for an entire career.

Steps to promote teaching in early-career faculty can enhance the synergy between research and teaching that contributes both to more relevant teaching and to more innovative research. Such programs as NSF's Faculty Early Career Development (CAREER) program[13] and the HHMI Professors program[14] help to bridge teaching and research and to support faculty members who excel in and integrate both.

Institutions can and should also support the development of good teachers. As discussed above, faculty development is a vital component, but generally helping to build institutional capacity should be a goal. That can include discussions of teaching and learning during faculty meetings, hosting speakers on education as part of department seminar series, offering certificate programs in undergraduate education for graduate students, designing new classroom spaces that support active learning, and providing opportunities for the development of new seminars and laboratories.

Those steps, taken together, can foster a culture of excellence in undergraduate education in which faculty, staff, administrators, and students work together to improve teaching and learning. Faculty who receive training in evidence-based methods and materials can be more effective teachers and promote enhanced student learning. The ultimate outcome should be well-prepared students who have the motivation and confidence to pursue their interests and careers of choice. Using research-based methods and supporting instruction that fosters these goals will help our universities to be leaders in undergraduate education.

---

[13]See <http://www.nsf.gov/career/> for more information about the NSF CAREER program.

[14]See <http://www.hhmi.org/research/professors/> for more information about the HHMI Professors program.

4

Breaking Down Silos in the University

Other chapters have noted the need for change in colleges of agriculture. This chapter will continue that theme, exploring how colleges of agriculture fit into the context of contemporary universities and meet the needs of students not only in the colleges themselves but throughout the university and beyond. It is clear that colleges of agriculture offer much to their universities and that universities in turn can and should provide expanded opportunities and resources to the colleges. To move forward, colleges of agriculture must always be aware of the need to change and collaborate. Only in this way will the colleges themselves ensure their continued relevance. In keeping with the subject of this report, it is important to emphasize that teaching and research should be mutually *supportive* rather than mutually *exclusive* activities. Some parts of traditional agriculture colleges have had only minor roles in undergraduate education beyond their own majors and little engagement in general education. It would be useful for all departments of agriculture colleges to be involved in undergraduate instruction more broadly.

The college of agriculture has many interests and activities in common with the rest of the university. Disciplines in the college of agriculture have strong intellectual connections with those in departments in many other parts of the university and should be seen as important contributors to the overall intellectual landscape of the larger institution. Much scholarship pursued by faculty in agriculture departments is in disciplines similar to those in other colleges in the university and uses equivalent and often identical techniques. Indeed, faculty members across the campus have the same or similar backgrounds and training. That many graduate programs cross the boundaries underscores the intellectual relationships.

At the same time, the college of agriculture has a culture of research-based service that makes it somewhat distinct from many other units in the university. A commitment to science-based problem-solving is important to students across the academy and should be actively included in the teaching efforts for undergraduate, graduate, and professional students to prepare the next generation of discipline-based specialists. Departments, faculty, and the college of agriculture share a basic goal of improving the human condition and the environment that we create and inhabit. The land-grant mission of instilling practical knowledge and understanding in a broad spectrum of students reinforces the need for the agriculture college to be seen as an important—even central—player in the liberal education of all students.

DESIRED QUALITIES OF GRADUATES

A concern for developing well-rounded students must be central to any consideration of college educational activities, and planning efforts should include plans for assessment. The following list enumerates many attributes that every education program should strive for. All students

- should acquire habits of disciplined learning, intellectual curiosity, and independence of mind;
- should think critically, follow trains of reasoning, engage in evidence-based reasoning, detect fallacies in arguments, discern unstated assumptions, interpret data, understand scientific approaches and recognize nonscientific arguments, and know how to construct, in speech or in writing, a sequence of logically connected and complex ideas;
- should develop essential competencies such as writing, interpersonal skills, quantitative and qualitative reasoning, and analytical and computational skills;
- should understand their own personal experiences more deeply and develop their capacity to empathize with others, especially those of different heritage, race, sex, or culture;
- should develop a sense of civic responsibility and ethical reflection and be prepared for responsible citizenship with an understanding of and strategies for dealing with such social issues as technology and society, the environment and the need for sustainability, multiculturalism, and the international dimensions of contemporary life;
- should recognize and anticipate the implications of actions, appreciating the societal impact of advances and activities;
- should become aware of some of the many ways in which contemporary life has been shaped and influenced by the intellectual and aesthetic traditions, moral and religious values, and economic and political structures

surrounding a field of study—in this case, agriculture, food, environment, and natural resources; and

- should increase their aesthetic sensitivity; improve their power of distinguishing what is well done from what is poorly done; and enhance their capacity to recognize a well-tuned sentence, a handsome building, an elegant proof, or a graceful move by a dancer or athlete.

Each of those qualities extends beyond specific disciplines. The committee believes that they should be integral parts of any curriculum or course of study.

PROVIDING A PROBLEM-SOLVING OUTLOOK TO THE BROADER COMMUNITY THROUGH EXTENSION

All agriculture colleges contain teaching and research in both basic and applied science. For many, basic-science research is explicitly used in the service of applied activities and yields solutions to real-world problems. Given the need for solutions to many problems of natural and human origin, harnessing science to find answers is both effective and timely. Agriculture colleges specifically and universities more generally are well placed to be leaders in problem-solving in the contemporary academy and should be poised to assume a central role in university-wide efforts involving both teaching and research.

The role of faculty traditionally includes teaching, research, and service, which, in many agriculture departments, means teaching, research, and extension. Agriculture faculties usually recognize all three missions; in contrast, many departments outside these colleges give much less, if any, official credit for activities beyond the university. Agriculture colleges often partition faculty efforts into two categories: research and teaching or research and extension. This has the unfortunate consequence of isolating some exciting and important research and extension efforts from the student community at large, as faculty often have significant engagement in either education or extension—but not both. (Chapter 5 includes an extensive discussion of involving undergraduates in outreach and extension activities.) It seems prudent for agriculture faculties to consider ways in which not only their broad array of disciplines but their approach of folding outreach and extension into a legitimate research-based activity could be made more accessible to the broader university community. University-wide undergraduate instruction in agriculture may be one solution.

PROVIDING COURSEWORK ELEMENTS BEYOND THE COLLEGE

In addition to offering agriculture courses to students throughout the campus, agriculture faculty could contribute to courses offered in other colleges. Agriculture and the disciplines studied in agriculture colleges can provide diverse, unique, and compelling examples and material for many natural-science and social-science courses. The National Research Council report *Bio2010: Transforming Undergraduate Education for Future Research Biologists* (NRC 2003a) makes the case that much of modern biology advances at the intersections of disciplines and that instruction should reveal and explore these intersections in more detail. Among its recommendations, the report proposes a modest solution: faculty in one field could contribute modules to courses in other fields. Faculty interests and the faculty themselves in virtually all departments in colleges of agriculture could enhance a wide variety of courses throughout the university.

To be sure, many communities advocate for the addition of their topics of interest to the curriculum. That is in part why the committee suggests focusing on modules that use agricultural examples to present content that is already being addressed in the other courses; that is, the modules would enable faculty in other disciplines to bring agriculture into the context of existing syllabi. For example, a genetics course could include examples about plant breeding, an engineering course could examine the development of agricultural technology, and a chemistry course could use examples from food sciences. Because of the importance of agriculture and its firm grounding in the natural and social sciences, the committee encourages universities, professional societies, government agencies, and others to include both agricultural and nonagricultural disciplines in discussions about curriculum at the institutional and national levels.[1]

CONNECTING WITH THE REST OF THE UNIVERSITY

In general, colleges of agriculture have faculty whose interests overlap with faculty outside of agriculture—throughout the life sciences, environment, and applied social sciences. Their disciplinary interests evolved from the needs of production agriculture and have expanded far beyond their original targets into topics of interest to faculty and students throughout the

---

[1] For example, the National Science Board has recommended the establishment of a National Council for STEM Education to "facilitate a strategy to define national STEM content guidelines" (NSB 2007).

university. If teaching and research programs can be organized around those topics, it might be easier to explore and develop cross-campus connections. Many research universities have found mechanisms for including faculty of different colleges in a single graduate program, but far fewer encourage or even allow this sort of collaboration for undergraduate instruction.

To facilitate enduring cross-college undergraduate teaching activities, college and university administrators need to find means of rewarding and supporting faculty who provide instruction to students from outside of their home colleges. At the department level, faculty must recognize—and treat as legitimate—this sort of cross-unit activity.

The committee believes that there are numerous opportunities for faculty from several departments and colleges to collaborate in courses that cover shared issues. In particular, there is often a significant amount of overlap at the introductory level where a single, well-designed course might serve the needs of agriculture, biology, chemistry, and other departments. One long-standing example is a course in world food problems at the University of Minnesota, which has been offered continuously since 1964 (Box 4-1).

BOX 4-1
The "World Food Problems" Course at the University of Minnesota

The University of Minnesota has been offering a multidisciplinary course that looks at problems and solutions affecting food production, storage, and use since 1964. Originally cross-listed in five departments, the course now involves faculty from three departments in two colleges and guest speakers from the campus and beyond.

Originally established as a capstone course for students working toward a minor in international agriculture, the course now enrolls graduate students from across the university. In fact, the student composition is one of the most diverse of any at the university with respect to major, background, and international status. Previous background in any of the disciplines is not required.

Presentations and discussions in the course introduce and discuss sometimes conflicting views on population control, use of technology, and the ethical and cultural values of the people in various countries. Emphasis is placed on the need for governments, international assistance agencies, international research and extension centers, and the business sector to assist in solving complex problems.

FACULTY RECRUITING

Another possible mechanism for promoting partnerships would be a strengthening of connections in faculty recruiting. One method would be to pool resources and offer joint appointments in which faculty have more than one departmental home; this requires careful planning and execution when junior tenure-track faculty are sought. Each department's expectations would have to be stated at the outset and reviewed often during the pre-tenure years. A more modest approach would be to provide—and recognize as legitimate—adjunct or secondary status in other departments with clear statements of tenure expectations. An even less dramatic strategy would be to include faculty from diverse departments in several colleges on search committees to encourage the recruitment of faculty with diverse interests that cross departmental lines. University and college leaders should be open to searches aimed at recruiting faculty who would serve students from beyond their own home departments. A description of a new interdisciplinary hiring program in sustainability that occurs largely outside of the departmental structure is described in Box 4-2.

BOX 4-2
Michigan Technological University's Strategic Faculty Hiring Initiative

Michigan Technological University has recently developed a mechanism for interdisciplinary hiring that is managed campus-wide instead of through individual departments (Jaschik 2008). The Strategic Faculty Hiring Initiative hired seven faculty members in 2008 to focus on the theme of sustainability and used more than 90 faculty members from throughout the institution in the process.

Each of the 230 applicants was evaluated by three reviewers who judged them on the basis of a variety of factors—including their contributions to sustainability studies—but not on their fit to particular departments. Only after the candidates were selected did the interdisciplinary faculty committee overseeing the process determine possible department homes. Those hired were promised evaluations—including tenure reviews, if necessary—that involve multiple departments.

The hiring process was much more open than a normal faculty search. Candidates were told the names of other final candidates, and the entire university community was invited to comment on the live and videotaped presentations by the candidates (Michigan Technological University 2008).

Additional information about the Strategic Faculty Hiring Initiative is available at <http://www.mtu.edu/sfhi/>.

UNIVERSITY-WIDE OFFERINGS

Some departments in agriculture colleges have faculty whose research spans more than one discipline. For example, many departments of natural resources house faculty who work on basic biological mechanisms, often keeping track of populations with the tools of molecular biology; other members of the same department study environmental systems with novel physical monitoring techniques and methods derived from basic chemistry; and still others articulate national and global policy issues and solutions. Similar collections of cross-cutting faculty interests can be found in plant-oriented and animal-oriented departments, food-science and nutrition-science departments, and departments derived from traditional agricultural studies, such as agronomy and dairy science. Those departments have nurtured, possibly more than departments in many other colleges, the currently fashionable concept of multidisciplinary and interdisciplinary research, often established to tackle specific, clearly stated problems. They present an important opportunity for further connections at the intersections of multiple disciplines. A powerful arena for this form of cooperative activity focused on addressing issues from diverse perspectives can be found in appropriately designed team-teaching efforts and is not limited to research.

NEW AREAS OF INSTRUCTION

Some departments in colleges of agriculture have recently broadened their focus and presented new opportunities in teaching. The following examples illustrate how changes in emphasis can create new and vital educational opportunities in a department and provide new student opportunities in the college and beyond. In planning such shifts, college leaders should explore new and changing needs for undergraduate and graduate training throughout the university.

Departments of agricultural engineering have existed in many institutions for years. The traditional view of outsiders may be that these units are concerned with milking machines, tractors and other machinery, and systems associated solely with production agriculture. Although there is still a clear need to solve problems related to production, agricultural-engineering departments have almost all moved on, often changing their names, usually by adding some aspect of bioengineering. Many of the departments have strong ties to engineering colleges, and many collaborative curricula have been developed in which students get the best of both worlds from the two colleges. Teams of students with diverse backgrounds, drawn from around the university, could become engaged in solving the compelling problems associated with agricultural and biological engineering. This type of problem-solving approach has been shown to be a highly effective teaching strategy that enhances student learning and engagement (see Chapter 3 for additional discussion). In addition, the application of theory and practice to real-world problems of immediate concern can be compelling "hooks" to engage students. In short, problem-based and inquiry-based courses aimed at undergraduates can often be effective recruiting tools to attract and retain the best students from around the university into new fields.

Many departments of agricultural economics have seen undergraduate enrollments decreasing. Most land-grant universities include business schools, so it is a challenge for the departments of agricultural economics to find a unique niche. This may be a good example of a unit that requires intensive and honest self-inspection to prevent atrophy or even extinction. Agricultural economics departments have always had strengths in applied economics with emphasis on empirical methods and risk management. There may be opportunities to join existing units beyond the agriculture college to contribute to other business-degree programs. Even simple measures, such as a department name change, suggest that such programs are beginning to expand their research, such as incorporating expertise in environmental, resource, development, and community management and economics. Cornell University is a good example (Box 4-3).

THE ROLE OF COLLEGES OF AGRICULTURE IN NURTURING LIBERAL EDUCATION

Overall needs in education must be stated in relatively simple terms; science and society are always changing, and no curriculum established today will be the most appropriate several years from now. However, fundamental attributes—including confidence, motivation, responsibility, effort, initiative, perseverance, caring, teamwork, common sense, problem-solving (critical thinking), and persuasion abilities—will always be important and can be mastered.

Overriding needs for an integrative point of view require a transformation of academic thinking and, in the process, a remaking of education at all levels. The land-grant university was founded on a sense of place, an integrated landscape with people in need of help. The environment, which is not a subject or a discipline or a commodity or a resource, can be used as an integrative theme—no discipline need be excluded.

BOX 4-3
Evolution of Agricultural Economics at Cornell University

The evolution of agricultural economics at Cornell University can provide insight into ways that academic departments can change over time to reflect current needs.

1909 The Department of Rural Economy and Department of Farm Crops and Farm Management are established.

1920s The departments merge to form the Department of Agricultural Economics and Farm Management and adds faculty in land economics, farm finance, marketing and cooperatives, and local government.

1940s World War II increases the demand for research and outreach related to food production. Following the war, agricultural marketing becomes a major focus.

1950s With a decline in the number of farms in New York State, the Food Distribution Program is established to build upon the department's programs in marketing and its relationships with food processors and retailers.

1960s The Department of Agricultural Economics establishes teaching and research programs in international agricultural development.

1970s With increasing demand, an undergraduate business program with an emphasis on food and agricultural industries becomes well established. Faculty are added in environmental and resource economics.

1980s The undergraduate curriculum continues to evolve from an initial focus on agricultural business to general business.

1990s The department changes its name to the Department of Agricultural, Resources, and Managerial Economics to reflect changing teaching, research, and outreach missions. The department's undergraduate business specializations are included in the accreditation review of Cornell's business degree programs.

2000s The department name is changed again to the Department of Applied Economics and Management. It now offers undergraduate specializations in agribusiness, applied economics, environmental and resource economics, and international development and trade as well as traditional business specializations in finance, marketing, accounting, and strategy.

SOURCE: Department of Applied Economics and Management, Cornell University, and Departmental Overview at <http://aem.cornell.edu/news>.

For over two centuries, educators have debated the "true nature" of liberal education. The debate has shifted in response to changes in the structure of knowledge, the social makeup of students and faculty, and society's expectations for undergraduate education. This report extends the historical debate by describing connections of disciplines and activities of colleges of agriculture to the contemporary liberal-education agenda.

Agriculture colleges have all too often viewed liberal education as the domain of liberal-arts colleges. We reject that view and affirm that colleges of agriculture have critical responsibilities for liberal education. Many of our recommendations build on experience: there is a rich and vital heritage in food and agricultural sciences on which to draw.

Although definitions of the fields of knowledge deemed essential to liberally educated citizens have changed, the concept of intellectual breadth has been constant. We emphasize the continuing importance of intellectual breadth by asserting that there are broad fields of knowledge that are associated with distinctive ways of knowing and with which every liberally educated person should be acquainted. The founding of the land-grant system almost 150 years ago occurred in an era in which most of the nation's citizens were intimately familiar with food, fiber, and natural-resources systems. Today's citizens are no less dependent on those systems, but they have far less first-hand experience with agriculture and are commonly so detached from the systems that they lack the knowledge needed to make informed personal and public decisions that affect the health and well-being of citizens and the natural world.

Maintaining intellectual breadth as an organizing principle of liberal education requires students to pursue in-depth study and to master particular bodies of knowledge and modes of inquiry. Only through in-depth study, as typically experienced in an undergraduate major, can students begin to grasp how knowledge is created and come to understand with certainty how knowledge furthers individual and social understanding.

Liberal education in its largest sense has to do with essential attitudes and qualities of mind, among them the capacity for critical thinking; openness to new ideas combined with independence of mind; curiosity about the social, cultural, and natural worlds in which we live; appreciation of the complexities of knowledge and tolerance of ambiguity; and a capacity for gaining perspective on one's own life through self-examination and the study of others.

Graduates of colleges of agriculture will need to be prepared to live in a rapidly changing world characterized by proliferating knowledge, an exploding capacity to create and transmit information, increasing global interdependence, and growing diversity in the nation's social and cultural life. Colleges of agriculture—because of their multiple roles in the creation, transformation, and transmission of knowledge; their history of addressing issues and concerns of diverse constituencies; and their commitment to addressing both domestic and international issues—are ideally situated to model new approaches to interdisciplinary and multidisciplinary teaching and learning in our universities.

As mentioned several times throughout this report, the committee has chosen not to propose a single model undergraduate curriculum; the decision as to what works best for a given institution will necessarily depend on individual strengths, missions, and resources. However, the committee provides here a vision of an overall undergraduate experience in the hope that it will be useful for institutional discussions:

- *The physical and biological sciences* will introduce students to the intellectual basis of experimental science, using the content and context of agriculture, food, the environment, and natural resources as the basis of courses that illustrate the connecting thread from basic to applied sciences for the benefit of science and nonscience majors.
- *History and the social sciences* are deeply imbedded in agriculture, food, the environment, and natural resources. History and the social sciences strive to understand the dynamic interplay between individuals and institutions, structures and processes, and ideas and events that characterize human behavior and complex societies, now and in the past. No other human activity has a longer history or greater social consequences than the pursuit of food, shelter, and natural resources and is at the center of human well-being.
- *The humanities and the arts* help to orient us to an extremely complex and elusive world by showing us the most compelling, expressive, and innovative forms and arguments through which people have tried to examine, symbolize, and discuss the human condition. Agriculture, food, the environment, and natural resources are at the root of the humanities and the arts. For students today, the excitement of encountering these efforts to understand ourselves and our history is not just "instruments to achieve a better job or become a richer nation." It is an indispensable prerequisite to a more satisfying, more luminous life, a life lived with intelligence and awareness rather than stumbled through in the dark.

5

Extending Beyond the University: External Partnerships to Effect Change

This report focuses on the need to effect change in undergraduate education in agriculture. Through improvements in instruction, assessment, and curricula, colleges and universities will be able to provide a relevant education in the context of the evolving food and fiber system for years to come. Effecting that change, however, is not limited to undergraduates or even to higher education institutions.

Many opportunities for intervention that will indirectly affect the number, training, and composition of students interested in undergraduate study in agriculture occur outside universities. Many kinds of intervention help to expose students to agriculture during their precollege years, including formal classroom activities in K–12 settings and academic enhancement programs. Others involve various types of informal education settings, from such extracurricular activities as the National FFA Organization and 4-H to activities organized by local gardening groups.

Stakeholders in undergraduate agricultural education include employers outside the education sector who are interested in the "products" of the nation's colleges and universities. Companies, public agencies, and other organizations that seek to hire college graduates well trained in agricultural disciplines have an obvious interest in improving education. Despite employers' concern for the quality of college graduates, they often have few connections to undergraduate institutions and often limited awareness of undergraduate curricula. There is a need for enhanced communication and collaboration because agriculture professionals may not be aware of the issues and constraints faced by academic institutions; conversely, faculty, students, and academic administrators may have little understanding of the needs of industry or other nonacademic employers.

This chapter describes a number of programs that involve partners from outside the university that will lead to improvements in undergraduate education in agriculture. The committee believes that partnerships are not just value-added opportunities but essential components of systemic reform of agriculture education. As will be discussed in Chapter 6, many stakeholder communities will need to participate in changing how agriculture is taught, learned, and perceived.

PARTNERSHIPS WITH K–12 AND PRECOLLEGE PROGRAMS

Almost all undergraduates enter college after graduating from the nation's K–12 education system. Therefore, one strategy for increasing the number and quality of students pursuing undergraduate study in food and agriculture is to encourage more students to pursue careers in agriculture before they reach college. Even when the immediate target audience is at the K–12 level, precollege programs may play an important role in affecting the number and preparation of future undergraduates.

Over the years, a number of highly successful K–12 and other precollege programs have provided students and teachers with firsthand knowledge of the broader educational and career opportunities in the agricultural sciences. Several of the most prominent such programs have been developed or supported, at least in part, by colleges and universities. For example, a number of colleges and universities provide teachers with innovative curriculum and teaching materials and provide research-based internships for students.

However, many colleges and universities seem slow to engage in the partnerships despite the effect that K–12 and precollege programs can have on students' educational and career choices. In part, that may be because higher education institutions are unaware of the types of programs that have been developed or because faculty receive little benefit from engaging in "recruitment activities." As discussed in Chapter 3, faculty rewards play an important role in faculty motivation.

The committee believes that higher education can play a more substantial role in outreach to high school and other precollege programs. Precollege programs, in particular, often involve engaging students in educational or scientific activities—common in a college setting—and giving them a taste of what a career in a field will entail. Sometimes, that is done by developing curricular materials or offering an agriculture-focused curriculum; more intensive initatives may have extracurricular or summer programs that bring students to college campuses for research and study. Agriculture colleges are well positioned to address each of those activities. Fostering "engaged learners" at an early stage helps to provide a framework for the concept of a lifetime of learning.

Some examples of K–12 and other precollege programs are discussed below; a few of them have been in operation for more than 30 years.

K–12 Curricular Programs

K–12 curricular programs provide valuable classroom resources to supplement and enhance an existing curriculum by increasing coverage of agriculture (see Box 5-1 for an example). Many of the programs provide materials that are reviewed, tested, and evaluated by teachers, content specialists, and curriculum experts for quality, appropriateness, and content accuracy. The materials are often aligned with national and state learning standards[1] and help classroom teachers and curriculum coordinators to understand how they can fit into a curriculum without a sacrifice of required content.

Several curricular programs have associated faculty-development activities in which K–12 teachers have the opportunity to learn more about the materials and to be trained in their use. Many also have state-level networks that provide continuing local support from volunteers or state-level coordinators. It is also common for the programs to have partners in a variety of sectors, often including business leaders and policy-makers. Although there are often some connections to colleges and universities, higher-education institutions are not especially well represented among the programs' partners; this suggests that there are additional opportunities for university faculty to be engaged in developing materials and in working with K–12 teachers in faculty development and implementation.

The federal government has recognized the value of connecting K–12 students to agriculture. Although the bulk of the National School Lunch Act deals with such issues as nutrition, it also includes provisions for linking schools, agricultural producers, parents, and other community stakeholders to help students to understand the source of their food (42 U.S.C. 1769). Many states also have established farm-to-school programs that link students to producers.[2]

---

[1]The predominant national standards include the *National Science Education Standards* (NRC 1996b) and *Benchmarks for Science Literacy* (AAAS 1993).

[2]See, for instance, the National Farm to School Program at <http://www.farmtoschool.org/>.

BOX 5-1
Agriculture in the Classroom

Agriculture in the Classroom (AITC) is a grassroots program coordinated by the U.S. Department of Agriculture (USDA); its goal is "to help students gain greater awareness of the role of agriculture in the economy and society, so that they become citizens who support wise agricultural policies." AITC is regarded as a flexible educational program designed to supplement and enhance teachers' existing curriculum by providing teaching materials, strategies, interactive exercises, helpful links, and awards for excellence in teaching about agriculture. AITC is carried out in each state, according to state needs and interest, by people who represent farm organizations, agribusiness, education, and government. USDA supports each state organization by helping to develop AITC programs, acting as a central clearinghouse for materials and information, encouraging USDA agencies to assist in the state programs, and coordinating with national organizations to increase awareness of agriculture in the nation's students.

Additional information about AITC is available at <http://www.agclassroom.org/>.

Urban Agricultural Education Programs

As the nation's population has become more urban and suburban, there has been a decline in the number of students who grew up on farms. The urban and suburban environments potentially have many highly qualified students who would be interested in pursuing careers in food and agriculture but have not been exposed to such opportunities. The concept of specialized urban agricultural education programs has been around for more than 50 years, most notably since the development of the W.B. Saul High School of Agricultural Sciences in Philadelphia, Pennsylvania (Esters and Bowen 2004). The last 20 years have seen increasing interest in educators in establishing urban agricultural education programs in other major cities. Agriculture-focused schools can now be found in some of the nation's largest cities and include the Chicago High School for Agricultural Sciences in Illinois and the Agricultural Food and Sciences Academy (AFSA) near St. Paul, Minnesota.

These public or charter schools prepare students for leadership and professional opportunities in the agricultural sciences. In addition to a standard college-preparatory curriculum, they typically offer a number of agriculture-related courses, including both science-based and business-based courses. They also place an emphasis on engaging students in their learning, using hands-on and experiential approaches and problem-solving related to agriculture. AFSA also engages its students in public outreach, helping to increase agricultural literacy in the Twin Cities urban population; this type of community engagement at the high-school level can serve as excellent preparation for extension activities once students get to college.

Summer High-School Enrichment Programs in Agriculture

In addition to formal K–12 school environments, a number of summer programs are designed to provide precollege students with exposure to careers in agriculture. One of the most successful is the intensive summer enrichment program offered by the Governor's School for Agricultural Sciences in a number of states, including Pennsylvania, Tennessee, and Virginia (see Box 5-2 for an example). Governor's schools offer several week-long summer academic experiences for high-achieving students and are generally on the campuses of state public academic institutions.

In part because they are on college campuses, such summer residential

BOX 5-2
Virginia Governor's School for Agricultural Sciences

The four-week Virginia Governor's School for Agricultural Sciences was started at Virginia Polytechnic Institute and State University (Virginia Tech) in 2004 with 52 students and has since grown to 92 students. Such organizations as the Virginia Farm Bureau and the Virginia Agribusiness Council recognized that an agricultural governor's school would be a tool to develop gifted and talented students' knowledge of the food and fiber system, recruit students to study agricultural sciences in higher education, and motivate them to pursue careers in the industry. The Department of Agricultural Extension Education in the Virginia Tech College of Agriculture and Life Sciences is the administrative body for the school, and the department's faculty and staff develop the curriculum and activities (Cannon et al. 2006).

Students selected to attend the school choose a major in agricultural economics, animal sciences, food science, natural resources, plant science, or veterinary medicine. Each student takes a course in each of the six fields of study and one specialized course in his or her major (Cannon et al. 2006). Students also take elective courses, such as communication and leadership, and participate in independent group projects, which allow students to conduct research on real-world problems related to agriculture in Virginia.

Additional information about the Virginia Governor's School for Agricultural Sciences is available at <http://www.gsa.vt.edu>.

programs provide some of the clearest connections between K–12 students and four-year institutions. For example, the Pennsylvania Governor's School for Agricultural Sciences involves about 70–100 faculty and staff from Pennsylvania State University's College of Agricultural Sciences each year. It is not uncommon for governor's school participants to choose to attend their state's college of agriculture, and attracting students seems to be a common goal of such programs.

Similar in some ways to governor's schools are high-school summer research programs. They provide students the opportunity to spend from a week to two months conducting research on a college campus. High-achieving high-school juniors and seniors are paired with faculty or graduate-student mentors. Many of these programs are targeted at members of under-represented minorities.

Even briefer, the Iowa Agricultural Youth Institute brings Iowa high-school sophomores, juniors, and seniors together for a four-day retreat on agricultural career opportunities and issues facing Iowa and U.S. agriculture. Students in the program have the opportunity to participate in such educational experiences as a team-building course, travel to the Iowa State Capitol, and a roundtable discussion with Iowa commodity representatives.

The committee believes that there are substantial opportunities for states and universities to expand the scope and size of these programs. States without agriculture-focused summer programs may wish to start them. They seem not only to help to expand the number of high-achieving students interested in agriculture but to help to connect high-school students with the state's colleges and universities. States that already have programs may wish to consider whether they can be expanded in size, inasmuch as such programs typically reach fewer than 100 students a year. Even without sponsorship from a governor's office, colleges and universities may be able to initiate similar programs on their own. In addition to educating and attracting students, such programs constitute an important way to connect university faculty with K–12 teachers.

There are also opportunities to incorporate agriculture into existing programs. For example, the Center for Talented Youth (CTY), run by Johns Hopkins University, enrolls over 10,000 gifted and talented students per year in summer programs at sites throughout the country.[3] Adding courses in agriculture to several of the CTY programs would expose a collection of some of the nation's best middle- and high-school students to the excitement and opportunities in agriculture.

[3]See <http://cty.jhu.edu/> for more information about CTY.

Youth-Enrichment Programs in Agriculture

In addition to formal curricula and academic programs, there are opportunities to provide K–12 students with exposure to agriculture and related fields through extracurricular youth enrichment programs, agricultural science clubs, and the like. Such programs can complement coursework and allow students to have a long-term engagement in learning about agricultural concepts.

Two of the most prominent such programs are 4-H and the National FFA Organization, both of which have connections to federal agencies: the U.S. Department of Agriculture for 4-H and the U.S. Department of Education for FFA. Each provides opportunities for young people across the country to be involved with an agriculture-focused national organization, to gain leadership skills, and to connect with scientists, practitioners, and other agriculture professionals.

The 4-H network, for example, claims to reach nearly 6.5 million young people through locations in all 50 states and territories and makes connections to higher education through programs at more than 100 land-grant institutions.[4] FFA, founded in 1928 as Future Farmers of America, reaches over 500,000 members 12–21 years old through over 7,000 local chapters.[5] More than one-third of FFA members live in urban and suburban areas, and there are chapters in 11 of the 20 largest cities in the country.

There are also programs that specifically expose minority-group students to educational and career opportunities in the agricultural sciences, including the precollege outreach program of the National Society of Minorities in Agriculture, Natural Resources and Related Sciences—Junior MANRRS—and the Retired Educators for Youth Agriculture Program, which bridges minority-group youth and agriculture professionals in Oklahoma.

In addition to programs focused on agriculture, several general youth-development programs include some exposure to and programming around agricultural issues, including the Boy Scouts of America, Girl Scouts USA, Boys & Girls Clubs of America, and Big Brothers Big Sisters.

Many of the messages in the report about the changing nature of agriculture also apply to the way that it is portrayed in youth-focused programs. These activities have the same responsibility as agriculture faculty to ensure that the treatment of agriculture in courses and curricula reflects the cutting edge and the increasing focus on issues such as sustainability and concern for the environment.

---

[4]See <http://www.4-h.org/> for more information about 4-H.

[5]See <http://www.ffa.org/> for more information about FFA.

PARTNERSHIPS BETWEEN ACADEMIC INSTITUTIONS

Academic institutions seem to exist largely in isolation from one another. Connections even within the same geographic area are often based on personal connections between individuals rather than institutionalized. Each institution may try to excel in everything rather than partner and choose to create stronger opportunities for all. Partnerships between academic institutions can take several forms, including building connections for students to move from one institution to another and establishing joint and multi-institutional programs that are stronger than any institution can do on its own.

Connecting Two- and Four-Year Institutions

It is increasingly common for students to enroll in community colleges instead of beginning their undergraduate study at four-year institutions; community colleges now enroll nearly half of all U.S. undergraduates, including 47% of black and 55% of Hispanic undergraduates.[6] To interest those students in possible careers in food and agriculture, it will be essential for community colleges to offer programs in agriculture and to facilitate the transfer of community-college students into four-year agricultural degree programs.

Many states are promoting transfer between two-year and four-year institutions to increase systemic efficiency and effectiveness in educating their citizens (Ignash and Townsend 2000). The most common type of collaborative effort among four-year institutions and community colleges has been the articulation agreement, a formal agreement that identifies the types of credits that transfer and the conditions under which transfer takes place (Kisker 2007; Zirkle et al. 2006). The committee believes that there are particular opportunities to extend articulation agreements with two-year institutions among the 1994 tribal land-grant colleges and other minority-serving institutions to provide opportunities for members of underrepresented minorities to advance their education. Articulated programs of study have several benefits, including ease of transition from one institution to another, articulated courses that may eliminate the coursework duplication

[6]From American Association of Community Colleges (AACC) analysis of January 2007 data from AACC, U.S. Department of Education, and College Board, accessed February 2008 <http://www2.aacc.nche.edu/research/index.htm>.

that some students experience as they move from one institution to another, and a reduction in educational expenses.

Most states operate articulation agreements under deregulated or regulated transfer systems. In a deregulated state system, individual institutions have the responsibility for establishing articulation agreements about which courses, programs, and degrees will transfer from one institution to another. In a more regulated system, the state may provide some general guidelines and incentives for institutions to develop the agreements; in a highly regulated system, a state may mandate that the associate of arts degree be accepted at all state institutions, as is the case in Florida (Ignash and Townsend 2000).

In one study, Ignash and Townsend (2000) found that 34 of 43 states had statewide articulation agreements. Fifteen of them had developed or improved existing agreements within the preceding five years—an indication of the attention that articulation and transfer policies have received from state higher-education agency officials, legislatures, colleges and universities, and the public in the last decade. In some states, the impetus to develop strong articulation agreements was a legislative mandate. Ignash and Townsend (2000) noted the need for improvements in developing articulation agreements for program majors and for the inclusion of private institutions in statewide agreements.

Articulation agreements are beginning to play a role particularly in teacher education: universities are strengthening partnerships with community colleges to prepare elementary-school and secondary-school teachers (Zirkle et al. 2006). Box 5-3 describes an articulation program in Ohio that addresses a shortage of business-education teachers, and Box 5-4 provides an example related to teacher education in Texas. Those efforts are meant both to address teacher shortages in subject-matter fields—such as mathematics, science, and agriculture—and to assist in the hiring of teachers who have diverse racial and ethnic backgrounds (Townsend and Ignash 2003).

Although articulation agreements have been touted as an essential first step in providing broad access to the baccalaureate degree (Ignash and Townsend 2000; Rifkin 2000), many scholars have argued that educators must move beyond articulation agreements to active collaboration with complementary institutions (Case 1999; Chatman 2001; DiMaria 1998). One type of partnership that has emerged in recent years is what Kisker (2007) has referred to as a *transfer partnership*—a collaboration between one or more community colleges and a bachelor's degree–granting institution for the purpose of increasing transfer and baccalaureate attainment for all or for a particular subset of students.

BOX 5-3
Articulation for Business-Education Teachers in Ohio

To address the shortage and diversity of business-education teachers in Ohio, Ohio State University (OSU) and Columbus State Community College (CSCC) recently developed an articulation program designed to allow a seamless transfer between the two institutions. The primary rationale for the development of the program focused on four points: the location of both institutions in Columbus, institutional missions that mention the need for community outreach and linkages, the OSU College of Education's goal of exploring ways to be on the cutting edge of new initiatives, and the opportunity for OSU to recruit a diverse student body into its teacher-education program from the student population of CSCC.

Preliminary results indicate that the OSU–CSCC articulation program has resulted in an innovative approach to addressing the shortage of teachers in business education. Its success can be ascribed, in part, to the inclusion of specific attributes characteristic of successful agreements, including taking the first two years of coursework at the community college, students' ability to complete most of the university general-education requirements at the community college, junior-class standing for students transferring to the university, and easy transfer and articulation policies to provide OSU credit for coursework taken at the community college (Zirkle et al. 2006).

BOX 5-4
Articulation for Teaching Education in Texas

Texas A&M University–Commerce (TAMUC) and Collin County Community College District (CCCCD) partnered to develop a program for articulated teacher education. CCCCD was the first community college in the country authorized to provide professional certification of teachers. TAMUC has a strong history in teacher education and sought to provide master's-level coursework in conjunction with the CCCCD teacher-certification program (Chambers et al. 2003). This alternative teacher-certification model established a university–community partnership designed to ameliorate the national shortage of qualified teachers. The TAMUC–CCCCD partnership provides a venue for people working toward certification through the community college to be awarded graduate experiential credit toward a master's degree that is not traditionally awarded to students taking courses at community colleges.

One essential element of the success of the partnership is a mutual commitment of each institution that outlines several criteria, such as enrollment requirements and use of classroom space and educational-technology equipment. Perhaps the most important effect of the TAMUC–CCCCD partnership is that it allows students to extend their education toward a master's degree while they are completing teacher-preparation courses at the community college (Chambers et al. 2003).

One partnership that has achieved success involves a large public research university in southern California and nine area community colleges (Kisker 2007). The partnership was established to develop a rigorous transfer-focused academic culture in each community college by addressing the persistent problems of weak academic preparation and inadequate academic counseling. Specific goals included increasing minority-group members' transfer to the university, using strategies that academically accelerate—rather than remediate—underprepared students, and promoting interaction between two-year and four-year faculty and discussion about preparing students for coursework at the university level (Kisker 2007). Partnership activities included several programs, such as implementing a rigorous theory-based tutoring model, accelerating community remedial sequences, and bringing two-year and four-year faculty together to discuss how they could arrange the community-college curriculum to facilitate student matriculation. Kisker (2007, p. 297) noted that "the utility of community college–university transfer partnerships is greater than simply increasing the number of students who move from one institution to another." In particular, transfer partnerships can raise students' awareness of the opportunities available to them after community college, assist in marketing and public-relations efforts, and create a culture of transfer on community-college campuses, especially among faculty.

As another example, Iowa's public and private four-year colleges and universities have historically had strong relationships with the state's community colleges (Blong and Bedell 1997). By the 1980s, community colleges and the three state universities[7] had signed articulation agreements that allowed any person who had earned an associate in arts degree at an Iowa community college to enter a state university with junior status in the college of liberal arts. Recently, Iowa State University and Iowa Valley Community College District (IVCCD) joined forces to make it even more convenient for IVCCD students to transfer to Iowa State. Through a joint admissions program known as the Admissions Partnership Program, IVCCD students who plan to pursue a bachelor's degree at Iowa State will receive special benefits to pave the way for academic success at both schools, including academic advising and career counseling, opportunities to participate in early orientation and registration before transfer to Iowa State, and guaranteed acceptance into a bachelor's degree program at Iowa State, provided that all college and program requirements are met at the time of transfer.

[7]Iowa State University, University of Iowa, and University of Northern Iowa.

Connecting Institution Types

Expanded partnerships may allow better integration of large research-intensive land-grant institutions with the 1890 historically black and 1994 tribal institutions[8] and with community colleges. Because 60% of tribal colleges have articulation agreements with local high schools, expanded partnerships could allow connections from the K–12 system to land-grant universities via tribal colleges. In fact, Kisker (2007, p. 299) argued that "community colleges occupy a unique position within a network of educational institutions that enable them to work with both high schools and 4-year universities." By instituting and publicizing transfer partnerships, especially partnerships that include all three educational sectors, two-year colleges can become the central agency to assure students a seamless transition from secondary school to college degree (James et al. 2001).

There are 32 tribal colleges and universities (TCUs) spanning 12 states. Several offer four-year degrees, although most remain two-year institutions that focus on certificate and associate degree programs (James et al. 2001). Key components of the TCU curriculum are cultural studies, community service, internships, and business training. Most TCUs seek to award transferable certification and maintain articulation agreements with four-year institutions to ensure that course credits can be transferred (Cole 2004). For example, the College of Menominee Nation in Wisconsin has articulation agreements with the University of Wisconsin at Stevens Point and Green Bay and with Wisconsin technical colleges in Wausau, Appleton, and Green Bay (American Indian College Fund 1996). In 1993, under the leadership of the Montana University System, 15 community colleges, tribal colleges, and other state-funded colleges and universities agreed on a core of 30 semester-hours that, if taken at one institution, could be applied as a block to the general-education requirements at another (Crofts 1997); the agreement was reached by a course-by-course identification of equivalence at the institutions.

Establishing Multi-institutional Centers of Excellence

Academic institutions may be able to do more with less by establishing multi-institution partnerships in which they work together on programs of common interest. The resulting consortia can offer a wider array of high-quality programs and opportunities than can a single institution alone. Such partnerships allow cost savings by diminishing the duplication of resources.

---

[8]See Chapter 2 for a discussion of the history and types of land-grant institutions.

BOX 5-5
Midwest Poultry Consortium

The Midwest Poultry Consortium was established in 1993, with the generation of the idea from the Midwest-United Egg Producers. The specific goals were to support and enhance poultry science programs in the Midwest, encourage students to enter poultry science, increase basic and applied research, and facilitate coordination in the poultry science community (Graves 1998).

Most relevant for this report is the consortium's Center of Excellence Program, which offers research-based education for students from 14 states in the Midwest and Florida. Although all courses are offered at the University of Wisconsin–Madison during two summer sessions, the faculty come from throughout the consortium and credits are transferred to the student's home university.

The program therefore provides access to students from a wide geographic area that might not be available at their individual campuses, and also provides access to laboratory training, industry field trips, and lectures and discussions with poultry science experts.

Additional information about the Midwest Poultry Consortium is available at <http://www.mwpoultry.org>.

They also allow for the growth of centers of excellence and foster opportunities for collaboration and exchange that extend beyond the consortia. As an example, the Midwest Poultry Consortium has created something akin to a "virtual poultry science department" that involves faculty and students from 14 states (Box 5-5). Washington State University and the University of Idaho have taken collaboration a step further, merging the two institutions' food science programs into a single Bi-State School of Food Science.[9]

INVOLVING UNDERGRADUATES IN OUTREACH AND EXTENSION

Land-grant institutions have a long history of outreach and extension in which university faculty and staff work with individuals and communities across the state to enhance agricultural knowledge and practice. However, those activities have largely been isolated from undergraduate education, and students rarely have the opportunity to participate despite longstanding agreement about the benefits that students gain from internships, practicums, service learning, and cooperative educational experiences—

[9]See <http://sfs.wsu.edu/> for more information about the Bi-State School of Food Science.

practical learning that has been shown to improve the quality of learning, increase student satisfaction, and enhance job placement. That disconnect indicates a need to encourage the involvement of undergraduates in outreach and extension.

The committee is enthusiastic about applied learning experiences for many reasons. They can challenge students to apply theory to practice, provide experience in solving complex problems, offer opportunities for communication to a variety of audiences, and build skills in negotiation and conflict resolution with diverse stakeholders. In addition, the experiences often provide a valuable service and link the university to the community. Involving undergraduates in extension is also a natural mechanism for integrating service learning and community engagement, which is becoming a field of concentration in many institutions (see Chapter 3 for additional discussion).

High-quality learning experiences in outreach and extension have the potential to recruit undergraduates to agriculture majors by giving them a glimpse of the diverse ways in which professionals contribute to community well-being. Facilitating the involvement of students in diverse disciplines will help to open their eyes to the exciting potential of careers in agriculture and natural resources.

To ensure high quality in practical learning, faculty must devote adequate time and resources to planning and oversight. Objectives, timelines, assignments, procedures, evaluation approaches, policies, and student expectations must be clear to both participating community partners and to students. Students must have accurate job descriptions and must not be assigned to menial work. The committee encourages opportunities for students to share their work through presentations or poster sessions on campus and in the community; nonmajor undergraduates, student newspaper reporters, faculty members, and community partners should be invited to the presentations.

Student internships and experiences in the extension service are advantageous because they provide a natural arena for applying theories learned in agriculture and natural-resources classes. In addition, they give students direct knowledge about career opportunities in extension (see Box 5-6 for an example from Florida).

Some opportunities in outreach are not associated with formal extension activities. An example in community-supported agriculture is discussed in Box 5-7.

BOX 5-6
Summer Internships in Extension at the University of Florida

For 6 years, the Institute of Food and Agricultural Sciences (IFAS) at the University of Florida in Gainesville has sponsored a summer internship program for 10 undergraduate students in county extension-service offices in response to proposals from county agents. Preference is given to minority-group students and those majoring in the College of Agricultural and Life Sciences at the University of Florida, but students in any accredited college or university in the state are eligible. Interns are asked to plan and teach programs at the local level under the supervision of an extension agent. The internship does not provide academic credit directly, but some students arrange to get credit by working with resident faculty advisers in their home institutions. Students are often placed in their own home counties, which makes housing arrangements less challenging. Seven former interns have been hired in permanent positions as county agents in Florida, and this provides at least anecdotal evidence that internships are an effective method of training and recruiting extension professionals.

Additional information about IFAS is available at <http://www.ifas.ufl.edu>.

PARTNERSHIPS WITH NONGOVERNMENTAL ORGANIZATIONS

There are a variety of nongovernmental organizations (NGOs) whose interests include agriculture; partnerships with these organizations offer opportunities for service learning and community engagement. Several are devoted to sustainable or organic farming or to fostering rural development. In fact, connecting with such groups can be a way to engage students directly with farmers (see Box 5-8 for an example). Others can connect students with those concerned with environmental impact, such as the Green Lands, Blue Waters Project described at the summit, which promotes multifunctional agriculture in the Upper Mississippi River Basin (see Box 5-9). A number of community-based independent organizations across the country foster students' interest in gardening. For example, Mixed Greens uses school vegetable gardens at ten public schools in Grand Rapids, Michigan, to teach urban youth about health, nutrition, agriculture, and the environment.[10] Growing Hope focuses on underresourced and disadvantaged populations in Ypsilanti, Michigan with school-based and community gardens.[11] These types of community-based organizations serve as important partners in

[10]See <http://www.mixedgreens.org/> for more information about Mixed Greens.
[11]See <http://growinghope.net/> for more information about Growing Hope.

BOX 5-7
Opportunities in Community-Supported Agriculture

Community Supported Agriculture (CSA) is a system of small-scale commercial gardeners and farmers. Shareholders pay in advance to cover costs of a farm or garden operation; in return, they receive a share of the farm's vegetables, flowers, fruit, herbs, milk, and meat products by way of weekly deliveries or pick-ups. CSAs are ideal for practical learning about production in a setting that values both high-quality food and high-quality care for the land, plants, and animals. They illustrate the characteristics of a small-scale closed market and can appeal to students' values and interests even if they are not majoring in agriculture or natural resources. Especially now, when more and more students in agriculture-related majors do not have any direct agrarian experience, CSAs can provide valuable experience and perspective to both majors and nonmajors.

The Cook Student Organic Farm at Rutgers University is operated as a CSA and is the largest organic farm managed by university students. The farm, founded in 1993, provides paid internships in the summer in which students learn about greenhouse operations, crop planning, pest and disease control, irrigation, post-harvest storage, soil building, fertilizer, composting, mulching, and weed control. Interns grow vegetables organically, gain experience in managing an operating farm, address issues of hunger in the community, and gain leadership training while they earn an income and raise their own food. The internship attracts a wide array of students; the farm's Web site (http://www.cook.rutgers.edu/~studentfarm/) shows interns majoring in nursing, public health, journalism, English, and natural resources. Students provide food for CSA shareholders and donate and deliver surplus produce to a local soup kitchen called Elijah's Promise.

Many universities with agriculture and natural-resources departments offer similar student farm experiences.

increasing public consciousness about agriculture and offer opportunities to engage precollege students in agriculture-related activities. In addition, some of the NGOs have sources of financial support beyond federal agencies, such as local foundations, local governments, and local businesses.

NGOs can also provide a number of opportunities that are discussed below with respect to employers. For example, faculty can look for opportunities to spend sabbaticals working at these organizations or serve in an advisory committee. Similarly, the leadership and staff at NGOs might be able to serve in various advisory capacities to academic institutions or to suggest problems and challenges that might serve as case studies in relevant classes. Internships and other student learning opportunities might be especially appropriate for NGOs: these organizations can get low-cost assistance

BOX 5-8
Connecting Farmers: Practical Farmers of Iowa

Those involved in production agriculture throughout the country are engaged in a number of activities that provide professional development for farmers. For example, Practical Farmers of Iowa (PFI) brings together over 700 members in Iowa and neighboring states to research, develop, and promote agricultural approaches that are ecologically sound, that enhance communities, and that have been found to be profitable.

Organized around sustainable agriculture, PFI organizes a number of programs and projects of interest to members in areas such as grazing clusters, developing niche pork markets, and improving horticulture through fruit and vegetable clusters. The organization not only fosters information sharing and community building, but can help promote science-based approaches to agriculture and help sustain family farms.

PFI has also been active in the educational arena, organizing a summer camp for youth and their families, offering a youth leadership program, and developing sustainable agriculture curricula for both elementary and high school students.

Additional information about PFI is available at <http://www.practicalfarmers.org/>.

BOX 5-9
The Green Lands, Blue Waters Project

The Green Lands, Blue Waters (GLBW) Project involves a partnership between more than a dozen nongovernmental organizations and several land-grant universities to support multifunctional agriculture in the Upper Mississippi River Basin that incorporates an increased number of perennial plants and other continuous living cover. GLBW incorporates such goals as sustainable grazing systems, use of perennial plants to obtain biofuels and oils, agroforestry, and wetland agroecology by working through an interdisciplinary, cross-sector collaboration.

The educational partnership involves formal coursework at affiliated institutions and summer internships in which undergraduate students in several disciplines are placed in a variety of enterprise development settings. Academic coursework at the University of Minnesota includes service-learning courses on the ecology of agricultural systems that incorporate systems thinking and an extensive service-learning project (Jordan et al. 2005). Another course offers a larger world-view challenge that explores the nexus of sustainable development, engagement, and professionalism; this course engages students collaboratively in considering the "Corn Belt" of 2036.

Additional information about the GLBW Project is available at <http://www.greenlandsbluewaters.org/>.

Summit presentation: Nicholas R. Jordan, *Professor of Agroecology, Department of Agronomy and Plant Genetics, University of Minnesota.*

on issues of concern while students can receive course credit for applying their classroom learning to real-world situations.

CONNECTIONS BETWEEN ACADEMIC INSTITUTIONS AND EMPLOYERS

Colleges of agriculture send many of their students to careers in industry, but students are often unaware of the full array of career options that await them once they leave the university. The committee sees many opportunities to develop the connections between academic institutions and employers. Some would directly affect student experiences, others would indirectly influence the undergraduate curriculum. The connections provide abundant benefits in enriching student experiences, enhancing career placement, and improving program quality. Partnerships at the faculty level can help faculty to understand the changing needs of industry, make connections with industry scientists, and learn real-world examples that can be taken back to the classroom.

Colleges must build true reciprocal partnerships and avoid viewing industry only as a source of funding, in-kind support, resources, and internships. Lasting relationships require that *both* parties benefit in a true reciprocal interaction. The committee encourages academic institutions to engage industry more fully in many of its activities, including asking for input on curricular decisions and for guidance on the kinds of educational programs that will best prepare their students for future careers.

Opportunities for Students

Agriculture and natural-resources programs and colleges are encouraged to devote adequate time and resources to developing internships and cooperative education programs in industry settings. Students and their supervisors need clear learning objectives, timelines, and definitions of deliverables, procedures, and policies. Students also need opportunities to showcase what they learn in internships to a wide audience, including to students in their own and other disciplines, faculty and administrators in a variety of departments and colleges, and partners outside the university. The benefits of poster sessions (or other mechanisms of sharing) are many and include student recruitment, résumé building, and enhancement of the reputation of the department or college. Boxes 5-10 and 5-11 describe two well-established partnerships between academic institutions and industry that provide opportunities for students to gain experience in the corporate world even before receiving their degrees.

BOX 5-10
Professional Practice at the Georgia Institute of Technology

The Division of Professional Practice at the Georgia Institute of Technology (Georgia Tech) has one of the oldest and largest optional cooperative-education programs in the nation. The program involves more than 3,000 student participants and 700 employers each year and is supported by a staff of 20. It is consistently ranked as a premier program. The division also houses a structured student internship program that includes an orientation program required of all participants.

The cooperative-education and internship programs both have carefully planned structures, policies, procedures, support systems, requirements for students and employers, and handbooks for students and employers. Student handbooks describe eligibility and requirements, policies, résumé writing, elements of a successful interview, the job and internship search process, and use of job-search tools. Employer handbooks describe benefits of the programs to participants and sponsors, requirements for employers, and the process of posting internship and cooperative-education positions and openings.

Georgia Tech places a high value on experiential learning and dedicates resources to provide a high-quality experience for all participants. Benefits to students include early career exploration, the ability to confirm career choices, developing skills in résumé writing and interviewing, honing job-search skills, beginning a professional network, earning a competitive wage while learning, and improving after-college job prospects.

Additional information about the division is available at <http://www.profpractice.gatech.edu>.

Summit Presentation: Thomas M. Akins, *Executive Director, Division of Professional Practice, Georgia Institute of Technology.*

The General Mills example in Box 5-11 illustrates the essential elements of strong partnerships and internships. Most important is that both partners benefit. The core academic programs gain interesting guest lecturers, bring successful graduates to the campus at the company's expense, and motivate students with the opportunity of well-paying and well-supervised summer internships that help them to compete for challenging first jobs. The company benefits by having access to high-quality students, building relationships with the students, and being able to hire outstanding young professionals who already know a lot about the company and can make wise decisions when they accept offers so that they are likely to remain with the company.

Although the committee endorses expanded opportunities for internships and other formal programs, more modest initiatives may meet with

BOX 5-11
Internships at General Mills

General Mills, in Minneapolis, Minnesota, identifies six core food-science programs in universities around the country on the basis of program quality. Core programs, which are highly ranked by General Mills scientists, provide a source of diverse students and have a track record of recruiting and retention success. The company designates an employee to serve as recruiting leader who is a graduate of the assigned institution and several more junior graduates who travel to the core campus each year. While on campus, company representatives attend career fairs, make classroom presentations, and interview applicants for internships and jobs.

General Mills has a well-developed internship program that seeks to identify high-quality candidates to take jobs after graduation. General Mills scientists compete to have interns work in their units by submitting proposals for intern-led problem-solving projects in their divisions or units. The best and most challenging proposals are chosen by a team of scientists and the company's human-resources department. The interns are assigned to technical units and have well-defined practical projects in those units when they arrive on the General Mills campus. An experienced manager supervises the assigned project, provides midcourse and summary performance appraisals, and offers regular coaching about personal and professional development.

General Mills uses a competence-based model for hiring and performance appraisal that also guides the choice and coaching of interns. Desired competences include judgment and problem-solving, energizing and developing people, delivering outstanding results, collaboration, adaptability and flexibility, technical excellence, leadership of innovation, and integrity and ethics.

success. College "career days" in which industry professionals visit with students and offer career advice can broaden the array of careers to which the students are exposed. Those intersections need not take place only on the college campus; opportunities for "job shadowing" and industry open houses can provide more information about the work of an agriculture professional in a single day than a week's worth of workshops.

Opportunities for University Faculty and Agriculture Professionals

University professors and food and agriculture professionals operate largely in different spheres. Although there are certainly some people who have moved between industry and academe, there are many benefits of increased permeability between various sectors. University faculty can gain increased insight into the corporate world, the kinds of problems that exist there and approaches to them, and firsthand experience with the opportunities that may be available to their students. Food and agriculture professionals can benefit from a more direct role in undergraduate and graduate curricula, and they have enormous expertise—and often different perspectives—to offer to individual students and to departments and institutions. In addition to the education benefits, fostering increased partnership between academic and nonacademic professionals also increases the likelihood of research collaboration. Intellectual property issues may pose a concern, especially with cutting-edge research, but the committee is hopeful that these issues can be addressed through general agreements and memoranda of understanding between academic institutions and their industrial partners. Box 5-12 describes a program at the Massachusetts Insti-

BOX 5-12
The Industrial Liaison Program at the Massachusetts Institute of Technology

The Massachusetts Institute of Technology (MIT) Industrial Liaison Program (ILP) is an example of partnerships between a university and a business outside the agriculture sector. Companies that pay a fee to join the ILP are assigned an industrial liaison officer (ILO) who has business experience and in-depth knowledge of MIT. The ILO is the direct contact for the company's managers, advocates for the company's needs, and serves as a liaison with MIT faculty and programs. Throughout the year, the ILO updates the company on MIT's activities, introduces MIT innovations and knowledge that could help the business, and takes other steps to meet the company's objectives.

Mars, Incorporated, is one corporation that has a partnership with the ILP. The company uses the partnership in various ways. For example, eight MIT doctoral students spent two months at the Mars technical center working on a project to optimize the company's manufacturing process and its economics on a global scale; and training classes provided for Mars managers by a faculty member at MIT's Sloan School of Management led to the adoption of a variety of new business techniques and new intellectual-property strategies. Mars research and development vice presidents noted that the ILO became a part of their research family, rather than an outsider, and that the ILO was a partner, not just an information provider.

The MIT ILP demonstrates several benefits that can accrue from academic–business partnerships. Students at all levels gain valuable experience in working on practical problems in real business settings, faculty members have opportunities to leverage their research and teaching, and member companies improve their processes and solve problems more quickly because they can access expertise and research results from a world-class research university.

Additional information about the ILP is available at <http://ilp-www.mit.edu>.

tute of Technology that involves students, faculty, and industry researchers in a multifaceted partnership.

INTERNATIONAL PARTNERSHIPS

Chapter 3 discusses the value of increasing the coverage of international perspectives for undergraduate students by both expanding opportunities for learning abroad and including global viewpoints in U.S. courses. Achieving these aims will require faculty members and graduate instructors who are knowledgeable about international issues and prepared to bring a variety of perspectives into their teaching.

International faculty exchanges and temporary international teaching assignments would increase the global perspective in both course content and research focus and should be encouraged. It will be important that such exchanges are rewarded in faculty promotion and tenure to reinforce the value that the institution puts on these experiences.

Programs could also be developed that would enable graduate students to spend a semester or year working and studying in another country. The international connections resulting from such exchanges will last for decades as graduate students launch their faculty careers with a personal understanding of the importance of international perspectives.

Unique approaches to funding and supporting globally focused programs should be developed. Universities should consider collaborations with foreign governments, and industry around the globe should be considered to make the programs lasting.

6

A Call for Change

This report has highlighted a number of challenges and opportunities that have the potential to transform undergraduate education in agriculture. In recognition that those opportunities will require action, this chapter outlines a number of essential recommendations whose implementation the committee believes is necessary for the future success of the agricultural sciences. The committee sees agriculture as uniquely positioned to respond to students' interest in making the world a better place and in responding to such important societal needs as food, health, environmental stewardship, sustainability, and energy security.

Implementing the recommendations described here not only will help to ensure the future of agriculture but may help to return many colleges of agriculture to their historical place at the heart of the university. Following through on the reforms called for in this report will require lasting commitment on the part of many stakeholders—students, faculty, departments, colleges, universities, industry and other employers, professional societies, farmers and farm organizations, commodity and interest groups, government and other funding agencies, environmental organizations and land trusts, food and environmental justices groups, science education organizations, community and other nongovernmental organizations, and others. All those players will need to participate in the conversation and play important roles in implementing the recommendations. The suggested interventions will require commitments of time, attention, and in some cases financial resources; the urgency and the need highlighted in Chapter 1 make the case for the critical nature of these investments.

On the surface, some of the recommendations may seem utilitarian and similar to those that have been made in past reports. Those who have been engaged in discussions about agricultural education for some time may see much that is familiar. But this report is directed to a much broader audience. Members of Congress, faculty outside of agriculture, and employers have not heard these ideas before, and the committee hopes that the messages will be compelling—and actionable—to this wider group of stakeholders beyond the college of agriculture and beyond the university.

Even if some of these ideas have been offered before, they have not been universally put into practice. The committee recognizes that many institutions have adopted *some* of the ideas in this report, but there are few institutions that have implemented *many*, and virtually none that have addressed *all*. The true power of these recommendations comes not in implementing one or even two ideas but in thinking about the entire system of agricultural education and in the synergistic combination of offering many different options. Although many of the individual ideas seem modest, the committee believes that they would be potentially transformative if universally adopted.

The committee has tried to provide advice about how stakeholders might respond to the recommendations by describing one or more sample implementations below each recommendation. These are meant to provide an example of how the ideas *might* be put into practice at different kinds of institutions, not a one-size-fits-all prescription on how they *should* be implemented. They are written to illustrate how the recommendations can be made real, but are not intended to be proscriptive or comprehensive nor will the particular examples be applicable to all institutions.

In addition to taking action, it is important that those implementing the recommendations described in this report simultaneously develop an evaluation and assessment strategy that will monitor the degree to which the interventions have been successful. The evaluations should be designed to provide formative feedback that will allow institutions and others to change their implementation strategy as the interventions are being implemented.

NEED FOR INSTITUTIONAL STRATEGIC PLANNING

The committee believes that all institutions offering undergraduate education in agriculture should engage in a period of conversation, self-study, and strategic planning—followed by putting the plan into action. The committee has chosen not to offer prescriptive recommendations for particular actions but instead to motivate attention to general focus areas and to provide examples of the kinds of steps that might be taken. The particular interventions that will respond to these recommendations will depend on the unique strengths, challenges, and circumstances faced by individual institutions, which can be addressed only by the institutions and their communities of stakeholders. In short, one size does not fit all in the specifics of implementing an objective.

As will be discussed several times in this chapter, strategic plans and conversations about the direction of undergraduate education in agriculture should be carried out in cooperation with a variety of stakeholders who have an interest in the undergraduate experience including those who employ graduates from agriculture colleges. That means not only students, faculty, and administrators from colleges of agriculture but also faculty from throughout the campus, professionals in teaching and learning, employers, local agricultural organizations, graduates, community members, and other interested parties. High-level academic administrators will need to be actively engaged in these discussions to be sure that campus leadership is committed to implementing the strategic plan and prepared to identify and commit the necessary resources.

RECOMMENDATION 1

Academic institutions offering undergraduate education in agriculture should engage in strategic planning to determine how they can best recruit, retain, and prepare the agriculture graduate of today and tomorrow. Conversations should involve a broad array of stakeholders with an interest in undergraduate agriculture education, including faculty in and outside agriculture colleges, current and former students, employers, disciplinary societies, commodity groups, local organizations focused on food and agriculture, farmers, and representatives of the public. Institutions should develop and implement a strategic plan within the next two years and to revisit that plan every three to five years thereafter.

Sample Implementation: Six months after the release of the report, one 1890 land-grant institution convened a steering committee of stakeholders from in and outside of the university to oversee a strategic planning process focused on undergraduate education in agriculture. The committee consisted of three faculty members from the School of Agriculture and Natural Sciences, a faculty member from each of the School of Business, the School of Health Studies, and Department of Social Sciences, the county superintendent of schools, and one representative each from a local seed company, a large farmer's cooperative, the State Department of Environmental Protection, and the State

Department of Rural Affairs. After a series of listening sessions with a group of stakeholders and discussions over the next 18 months, the plan was developed and refined, even despite the retirement of a key senior administrator at the university. Two years after the report, the plan is fully implemented, and the institution has formalized a process for regular review and amendment.

Strategic planning should be the beginning of an extended and ongoing process of change, evaluation, and adaptation. Implementation will need to follow the ideas, and pilot-testing and continual assessment used to refine and improve new programs and policies. The committee emphasizes that action and implementation are necessary steps for achieving the goals of this recommendation and encourages academic institutions to include timelines for implementation as formal parts of their strategic plans.

The committee reinforces that the stakeholders brought into discussions of undergraduate education in agriculture should be broader than those who have traditionally been involved. Faculty, students, and commodity groups should continue to be integral participants, but institutions should think broadly and include a more inclusive group of stakeholders in and outside the university than have been engaged previously at many institutions.

AGRICULTURE ACROSS THE CURRICULUM

One of the most important actions that institutions can take to enhance student interest in agriculture is to increase agricultural literacy. That means helping students understand such issues as where their food comes from and the role of agricultural products in energy production. It also means demonstrating that 21st-century agriculture means much more than farming.

Among the ways that more students can be exposed to agricultural topics are the incorporation of agriculture examples in courses outside agriculture and the offering of team-taught and interdepartmental introductory courses that serve students in a variety of majors. More radically, institutions may wish to consider whether the current organization of their natural and social science and engineering disciplines in and outside agriculture is most appropriate for today's research and education needs. Although the committee believes that agriculture colleges have a unique and continuing role, it may be appropriate for institutions to consider the organizational structure that is most appropriate for their own setting, as many institutions have already done.

RECOMMENDATION 2

Academic institutions should take steps to broaden the treatment of agriculture in the overall undergraduate curriculum. In particular, faculty in colleges of agriculture should work with colleagues throughout the institution to develop and teach joint introductory courses that serve multiple populations. Agriculture faculty should work with colleagues to incorporate agricultural examples and topics into courses throughout the institution.

Sample Implementation: The faculty at one of the nation's largest agriculture colleges decided that cross-disciplinary education was important and committed that each department in the college would offer at least one introductory course that is cross-listed with a department outside of the agriculture college. They sought support for this idea from the Provost, who provided a small amount of course development funds that enabled faculty across the campus to develop courses that fulfilled the curriculum requirements in their respective departments. The revamped series of introductory courses now enroll students from throughout the university and integrate agriculture with courses in several other colleges, including the College of Life and Environmental Sciences, the College of Humanities and Social Sciences, the College of Public Health, and the College of Business and Finance.

Sample Implementation: The provost at a non-land-grant institution decided to hold a meeting involving all of the faculty teaching introductory courses in science, technology, engineering, agriculture, and mathematics in the next semester. This meeting, which actually became a monthly conversation, helped to foster communication between the courses taught concurrently and enabled faculty to share their syllabi and suggest ways that the courses might be effectively integrated. Social science and humanities faculty are preparing similar coordination for their disciplines.

The committee further encourages agriculture courses to take advantage of research in student learning and to draw on real-world examples, engage students actively, and be informed by agricultural science and practice from a variety of viewpoints.

The committee hopes that interdepartmental connections extend far beyond course content and include a greater number of joint faculty appointments, interdisciplinary research and education centers, and structures for collaboration. The close methodological and content connections between disciplines in colleges of agriculture and throughout the university—in colleges of arts and sciences, education, medicine, and engineering, among others—demand that faculty communicate more directly and collaborate more often; it will often be necessary to break down administrative barriers to facilitate such interactions.

CHANGES IN HOW STUDENTS LEARN

During an undergraduate education, students should master a variety of transferable skills in addition to content knowledge. Employers value the skills at least as much as book learning. Communication, teamwork, decision-making, critical thinking, and management should be emphasized and made important parts of the curriculum. Rather than create new courses, the committee recommends that institutions integrate these experiences into existing courses so that students have opportunities to speak and write, to work together, and to lead and manage as part of the activities in their "standard courses."

Students should also have opportunities to engage in a variety of experiences that help to make the content knowledge come alive, including undergraduate research, internships and other extra-institutional programs, international experiences, and participation in service learning and in extension and outreach. The ability to connect undergraduate education and extension is an opportunity unique to colleges of agriculture; it not only expands the sphere of institutional and statewide extension and outreach but provides a chance for undergraduate students to give back to their communities and become spokespeople for agriculture.

RECOMMENDATION 3
Academic institutions should broaden the undergraduate student experience so that it will integrate:

- **numerous opportunities to develop a variety of transferable skills, including communication, teamwork, and management;**
- **the opportunity to participate in undergraduate research;**
- **the opportunity to participate in outreach and extension;**
- **the opportunity to participate in internships and other programs that provide experiences beyond the institution; and**
- **exposure to international perspectives, including targeted learning-abroad programs and international perspectives in existing courses.**

Sample Implementation: The College of Agriculture at a land-grant institution established a committee of faculty, students, and employers to develop a list of skills and competences that all students should have upon graduating. The list explicitly detailed how these skills were incorporated into its undergraduate majors or how they could be included by offering additional experiences. Two faculty members requested supplements that would support undergraduate research experiences in conjunction with extension. They received matching funds from the state soybean council to organize studies involving undergraduate students and farmers in identifying best practices for reducing run-off.

The committee recognizes that not all students will choose to participate extensively in all those activities, but every undergraduate should be exposed to them and have the opportunity to explore chosen ones in depth.

Providing such opportunities will require resources, but several can be provided at relatively low cost. In some cases, public and private funding agencies may need to provide new awards or to extend existing programs to new activities. In other cases, agencies might expand the use of supplements to existing awards to support specific educational aims; for example, the National Science Foundation (NSF) offers supplements to foundation-funded research projects to support undergraduate research experiences. Even without increased extramural funding, however, the committee urges universities to prioritize these experiences and to redirect institutional resources to support them.

As will be discussed below, some of the experiences might be made available to students through partnerships with companies and other organizations outside the university. Such opportunities as internships, cooperative education programs, and service learning can also help students to develop transferable skills, conduct research, and gain exposure to a wide variety of viewpoints and ideas.

CHANGES IN HOW FACULTY TEACH

The scholarship of teaching and learning has developed substantially over the last several decades. As outlined in Chapter 3, the consensus of the research is that students learn more when they are actively engaged and have the opportunity to consider real-world situations and examples. Nevertheless, universities still tend to use an outmoded method of teaching in which lecturing is the norm and the focus on facts is predominant. Many classes fail to engage students or to take advantage of the research in how people learn.

In general, university faculty do not receive much training in effective teaching, nor are they exposed to research on student learning; faculty in agriculture are no exception. Therefore, it will be necessary to provide opportunities for faculty to learn about the research on how people learn and to have access to resources to implement course and curricular changes. A variety of stakeholders will need to devote attention and resources to faculty development both in the short term and on a continual basis. The committee especially encourages graduate programs to build those topics and competences into training for the next generation of faculty.

Faculty will need access to professional-development opportunities and to the resources necessary for implementing effective instructional strategies. Educational innovation is generally much less expensive than investment in research, but it is not free. In fact, time may be a more precious resource than money for many faculty: time to develop new courses, redesign curricula, and identify, adapt, or create the necessary teaching materials.

RECOMMENDATION 4
Several actions are necessary to prepare faculty to teach in the most effective ways and to develop new courses and curricula:

- **Academic institutions, professional societies, and funding agencies should promote and support ongoing faculty-development activities at the institutional, local, regional, and national levels. Particular attention should be paid to preparing the next generation of faculty by providing appropriate training to graduate students and postdoctoral researchers. Moreover, academic institutions should take steps to ensure that the responsibility for faculty development rests not with individual faculty members but with departments, colleges, and institutions.**
- **Academic institutions and funding agencies should leverage existing resources or provide additional resources to support the development of new courses, curricula, and teaching materials. Among the needed resources are faculty release time, support for teaching assistants, attendance at education-focused workshops, and use of education materials and technologies.**

Sample Implementation: One institution restructured their resources for professional development to enable each faculty member teaching undergraduate courses to attend at least one education-focused *workshop per year. The dean of the college of agriculture committed to provide $5,000 in startup funds for novel educational endeavors. One junior faculty member used these funds to support a research study to develop and assess the effectiveness of an activity to teach a difficult aspect of plant biology; the study was subsequently published as a peer-reviewed article in the* Journal of National Resources and Life Sciences Education *and presented at the annual Plant Biology meeting.*

Sample Implementation: An agriculture college restructured its graduate curriculum to include a course in teaching and learning within agriculture as part of its core curriculum. In preparation for the course, the college sent two faculty members and two graduate students to a national meeting on enhancing the preparation of graduate students for careers in teaching and invited representatives from two institutions that have such a course to give a college-wide seminar and meet with faculty and students. The course, which is also available to postdoctoral researchers and to faculty, provides an overview of practical education, exposes students to teaching pedagogies and resources, and provides a forum for discussion of educational issues. The course has become part of the training for graduate teaching assistants (TAs), and TAs are asked to incorporate what they learn into their own classroom practice.

Many colleges and universities have developed centers for teaching and learning and have professional staff trained to provide support for high-quality teaching. Such centers are an ideal venue for programming and support for faculty, graduate students, and postdoctoral researchers in teaching. Institutions should look for opportunities to expand and enhance the services provided by such centers or to establish them if they do not already exist.

Institutions are encouraged to involve graduate students, postdoctoral researchers, and advanced undergraduates in developing educational materials and fostering excellence in teaching and learning. In addition to providing additional expertise devoted to improving education, the entire educational system benefits by engaging these potential future faculty members in thinking about teaching and learning early in their careers.

The committee notes that many of the issues related to faculty development also apply to teachers at the K–12 level. For example, a wealth of resources is available to K–12 teachers (such as those described in Chapter 5), but many teachers are unaware of them.

SUPPORTING THE VALUE OF TEACHING AND LEARNING

At the Leadership Summit, it was strongly expressed that achievements in teaching are rarely rewarded in substantive ways and that faculty were thus prompted to focus their attention elsewhere. That poses a particular challenge to the implementation of the recommendations in this report inasmuch as effecting change in undergraduate agriculture education will require attention to teaching and learning. Although a full vetting of tenure and promotion criteria and institutional priorities is well beyond the scope of this report, the committee offers several suggestions of actions that it believes are essential for improving undergraduate education in agriculture.

RECOMMENDATION 5
Several stakeholders should take tangible steps to recognize and support exemplary undergraduate teaching and related activities:

- **Academic institutions should enhance institutional rewards for high-quality teaching, curriculum development, mentoring and other efforts to improve student learning, including rigorous consideration in hiring, tenure, and promotion. Academic institutions should also implement new tenure-track faculty appointments that emphasize teaching and education research in the discipline.**
- **Funding agencies should support and reward excellence in teaching in both education and research grants. Such models as the National Science Foundation's "broader-impacts criterion" should be considered by other agencies.**
- **Professional societies should raise the profile of teaching in the disciplines. That may include offering support and rewards for undergraduate teaching and sponsoring education sessions and speakers at society meetings, workshops on teaching and learning, education-focused articles in society publications, and efforts to facilitate the development and dissemination of teaching materials.**

Sample Implementation: The faculty senate at one institution coordinated a review of tenure and promotion criteria, developing a set of rigorous criteria that enabled teaching quality, measures of student learning, and the level of faculty engagement with educational activities to be explicitly considered for hiring, tenure, and promotion decisions. The criteria developed also include methods of evaluation for measuring each of these qualities and assessing student learning and the effectiveness of instruction without overreliance on traditional student evaluations.

Sample Implementation: A major private funder of agricultural research began to require grant applicants to explain how their research would impact undergraduate education or how it would be incorporated into public outreach and extension activities. Applicants who wish to receive this funding must, therefore, commit to educational activities, along with evaluation of their impact and success. The sponsor organized a regional workshop of its grantees so that they might share their experiences and results with each other.

Sample Implementation: The board of a major professional research society in the agricultural sciences voted to enhance the profile of education within the discipline. Within a year, they committed to sponsor at least one education-focused plenary speaker at the society's annual and regional meetings and publish at least one education-focused article in each issue of the major research journal published by the society.

INCREASING CONNECTIONS BETWEEN INSTITUTIONS

Many colleges and universities offer programs in agriculture, but they tend to exist in isolation, with few connections between institutions even in the same geographic area. Moreover, community and tribal colleges are playing an increasingly important role in undergraduate education, enrolling large numbers of students and especially high percentages of members of groups traditionally underrepresented in four-year colleges. But there are few pathways for those students to pursue agricultural careers. Similarly, there are opportunities for colleges of agriculture to work with other, often smaller institutions to develop and enhance agriculture programs.

RECOMMENDATION 6
Academic institutions offering teaching and learning opportunities in food and agriculture should enhance connections with each other to support and develop new opportunities and student pathways. In particular, four-year colleges and universities should further develop their connections with community colleges and with 1890 and 1994 land-grant institutions. In addition, four-year institutions should work with other institutions to establish and support joint programs and

courses relevant to agriculture and develop pathways for students pursuing agricultural careers.

Sample Implementation: Four months after the release of this report, a major land-grant institution organized a meeting of all academic institutions within 200 miles that offer undergraduate instruction in agriculture. This group included several community colleges as well as 1890 and 1994 land-grant institutions. The meeting resulted in a commitment to develop cross-registration and articulation agreements to facilitate student exchange. A multi-institution faculty committee has also begun establishing a regional center of excellence in a field of agriculture relevant to the region, with support from USDA; when up and running, the center will offer both undergraduate and graduate instruction available to students at any of the institutions and will create a locus for research in that field.

Articulation agreements and transfer partnerships should be developed between two- and four-year institutions when appropriate—but connections should not be limited to those arrangements. Institutions may wish to develop multi-institution programs, share resources, allow easy exchange of faculty and students, and generally work together to support and promote initiatives of common interest.

Partnerships should exist without regard to an institution's official status as a land-grant institution but be based on common purpose and goals.

INCREASING CONNECTIONS WITH PRECOLLEGE SETTINGS

Reform of the role and perception of agriculture is a challenge far beyond the scope of this report, but it is clear that action in this area cannot occur solely in institutions of higher education. The committee believes that there are many opportunities to develop K–12 students' interest in agriculture, including formal academic programs and extracurricular programs, such as 4-H and National FFA. Higher-education institutions have a particular capacity to effect change in K–12 settings and a responsibility to lead.

RECOMMENDATION 7

Colleges and universities should reach out to elementary-school and secondary-school students and teachers to expose students to agricultural topics and generate interest in agricultural careers. Although the specific partnerships will differ from institution to institution, programs

that might be considered include agriculture-based high schools, urban agricultural education programs, and summer high-school or youth enrichments programs in agriculture. In addition to formal partnerships and academic programs, colleges and universities should explore partnerships with youth-focused programs, such as 4-H, National FFA, and scouting programs.

Sample Implementation: Four months after the release of the report, a non-land-grant college of agriculture called a meeting of the regional K–12 school systems as well as area chapters of agriculture-focused youth and community programs. One outgrowth of the meeting was the initiation of a program for undergraduate and graduate students to spend two days per month working with middle- and high-school courses in agriculture. Several university students also signed up to be mentors to students in the Boy Scouts interested in agriculture.

Sample Implementation: In one western state, the state board of education put out a call for proposals to the state's public institutions, asking them to propose programs in food and agriculture for secondary school students. The state made two awards: one recipient has established a Governor's School in food and agriculture that offers two four-week sessions each summer; the other recipient has developed a series of day-long activities that are offered to high-school classes surrounding its urban location.

INCREASED PERMEABILITY BETWEEN ACADEMIC INSTITUTIONS AND EMPLOYERS

Discussions at the Leadership Summit and elsewhere testify that academe and industry operate in largely distinct spheres, although industry is a major employer of food and agriculture graduates. Moreover, many employers have little understanding of how colleges and universities are organized, and academe has little understanding of needs outside the academic sector. Although a number of universities have long-standing partnerships with particular industries or corporations, there are many opportunities to expand such collaborations to a wider array of institutions, companies, and sectors.

To reduce the "silo effect," the committee offers a multipart recommendation to enhance communication and coordination between academe and employers at different levels. Each of the elements in the recommendation is meant to provide a mutually beneficial relationship. For example, students benefit from such activities as internships and cooperative education programs to gain real-world work experiences, and industry gains an opportunity to recruit and attract talented young people and hire workers who already have experience working in the company.

RECOMMENDATION 8

Stakeholders in academe and other sectors should develop partnerships that will facilitate enhanced communication and coordination with respect to the education of students in food and agriculture. The partnerships should include the following elements:

- **Academic institutions should include representatives of industry and other employers on visiting committees, on advisory boards, and in strategic planning. Companies should include academic faculty on their advisory committees.**
- **Exchange programs should be developed that enable food and agriculture professionals to spend semesters teaching and working at academic institutions and enable faculty to spend sabbaticals working outside of academe.**
- **Opportunities for students to work in nonacademic settings should be developed and greatly expanded. Programs might include internships, cooperative education programs, summer opportunities, mentoring and career programs, job shadowing, and other experiences.**

Sample Implementation: A regional agricultural business consortium partnered with a local college of agriculture to convene a meeting of area companies, academic institutions, and nongovernmental organizations (NGOs) with a stake in food and agriculture. As a result of the meeting, the business consortium agreed to coordinate a student internship program that would enable a cohort of students each semester to do an internship at one of the companies or local NGOs.

Sample Implementation: A national organization representing universities and one representing companies in food and agriculture partnered to establish a clearinghouse of opportunities for sabbatical research in industry and institutions willing to offer temporary visiting professorships for industry professionals. Representatives from the two organizations also developed template intellectual property policies that facilitate the exchange of people and information.

These opportunities need not be limited to large food and agriculture companies but could incorporate a wide range of employment sectors from small family farms to NGOs. The committee hopes that such collaborative opportunities will have important secondary benefits. For example, closer connections between academe and industry may encourage industry to call on academe for assistance in solving industrial challenges; such questions may serve as case studies in undergraduate classes and provide opportunities for undergraduate research.

ACCOUNTABILITY AND COMPLIANCE

In order to provide a strong incentive for implementation, the committee has developed a "checklist" of items that should be used by any individual or group conducting a review of a program, curriculum, department, college, or institution (Appendix E). Although the committee does not have the authority to enforce specific competences, it hopes that these elements will inform the establishment of review criteria and accreditation standards at all levels and in a wide variety of settings.

RECOMMENDATION 9
Organizations and individuals conducting reviews related to undergraduate education in agriculture should incorporate the elements discussed in this report (summarized in Appendix E) to guide their decisions and reports. This includes accreditation, review of grant proposals, department and other institutional reviews, and other venues.

Sample Implementation: Regional accrediting bodies include the list of questions in Appendix E as a recommendation for the institutional self-study as well as the external accreditation review.

Sample Implementation: An organization representing small agriculture companies decided to prepare a list of the skills and competences that they are looking for in hiring college graduates. The organization not only distributes the list to all agriculture-focused colleges but commits to including a session at its annual meeting every four years to review and refine the list.

For example, the U.S. Department of Agriculture (USDA) might incorporate more specific elements into the evaluation criteria for the review of its programs including—but not limited to—the Higher Education Challenge Grants Program.[1] USDA might also develop workshops for its staff that provide additional context and background for these issues. Accreditation bodies within the United States could use these elements to develop a specific set of benchmarks that institutions might be asked to meet to receive accreditation. External review and visiting committees might ask institutions and programs to meet the standards called for in this report. Peer-review panels might use the elements in Appendix E as goals that submitted grant proposals should seek to achieve. Professional societies could use these elements to guide discussions within disciplines and to make decisions of organizational priorities based upon those elements. The Association of Public and Land-grant Universities (APLU) can use the elements in this report to guide the content of teaching workshops and discussions among the Academic Programs Section.

The committee expects that monitoring implementation and change will itself become a topic for research and evaluation. Faculty and graduate students in agriculture education programs may see this as a fruitful area for long-term study, tracking change and determining factors that contribute to institutional change and effective implementation.

IMPLEMENTING CHANGE

The recommendations offered above refer to various stakeholders that will need to take action. Although it can be easy for one party to see a challenge as someone else's responsibility, the committee emphasizes that each of the many stakeholders has a role in and responsibility for improving undergraduate education in agriculture. For example, if employers want better-prepared graduates, they need to be part of the solution. If colleges of agriculture want students to understand that "agriculture" does not equal "farming," they need to reach students from throughout the university and the general public. If universities want to retain more students in agriculture majors, they need to foster teaching and learning that promotes student learning and addresses student interests. If agriculture is to be seen as

[1]Although the Higher Education Challenge Grants Program solicitation includes several priority need areas—including curricula design and materials development, faculty preparation and enhancement for teaching, instruction delivery systems, student experiential learning, and student recruitment and retention—the current *evaluation* criteria are quite vague. See <http://www.csrees.usda.gov/funding/rfas/pdfs/08_hep_challenge.pdf>.

science-based, it needs to take its place among other science, technology, engineering, and mathematics disciplines. The committee hopes that academic institutions, food and agriculture employers, government agencies, professional societies, and others will take the recommendations in this report seriously and implement changes to improve undergraduate education on individual campuses.

The Role of Students

As the prime "consumers" of education, students will be most directly affected by implementation of the recommendations in this report. Although none of the recommendations explicitly calls for action by students, the committee believes that students have a responsibility to become educated consumers and to be advocates for their own education. Students are encouraged to make the kinds of connections that are described in this report, enrolling in a variety of courses and taking full advantage of the opportunities they are given. Students should ask for and pursue the kinds of experiences that will serve their professional and personal interests, prepare them for a wide array of careers, and provide them with a valuable undergraduate experience.

This report is more likely to make it into the hands of faculty and administrators than into the hands of individual students, and the committee calls on colleges and universities to help students to fulfill their responsibilities. That is, we hope that academic institutions pass along the committee's encouragement to their students and engage undergraduate and graduate students as full participants in discussions about teaching and learning.

The Role of Faculty

Many of the recommendations in this report are focused on the classroom—what is taught and how. Thus, faculty members make up one of the primary audiences for this report and should be intimately involved in discussions about how to implement its recommendations. Faculty have primary responsibility for what and how material is taught, so they should pay particular attention to the discussions about course content and pedagogy. They can lead by example in devoting themselves to high-quality teaching in their departments, disciplines, and institutions and in recruiting and supporting colleagues who demonstrate a strong commitment to education. Faculty also make up departments, colleges, and universities and will need to be committed to the changes that these administrative structures seek to implement.

The Role of Departments

The academic department is often the most crucial level of organization in a university setting. Faculty appointments, promotion and tenure, undergraduate majors, graduate programs, credit for teaching, and even recovery of indirect costs are often tied to departments. The role and size of academic departments provide an excellent locus for reform of undergraduate education and for recognizing the scholarship of teaching. In fact, the committee hopes that *departments* will collectively take on the responsibility for teaching and learning, not relying on the good will of individual faculty. In addition, departments have the opportunity to work together on administrative and content issues to reduce barriers to cross-department offerings and provide students with cohesive undergraduate experiences.

The Role of Colleges of Agriculture

As described in Chapter 4, colleges that include agricultural disciplines have undergone extensive evolution and transformation, often incorporating such additional fields of inquiry as natural resources, environmental sciences, and life sciences in addition to traditional agricultural disciplines. As the home of agriculture, they have the most at stake and the most to gain from implementation of the committee's recommendations and from taking the ideas presented in this report seriously. Therefore, it is essential that agriculture colleges play a leading role in changes that may ultimately spread throughout the university, such as revisiting promotion and tenure policies, establishing a student-centered curriculum, and providing opportunities for students to engage with the wider community as part of their education.

Discussions should occur within and between colleges of agriculture and should involve faculty, undergraduate and graduate students, and staff. The APLU Academic Programs Section has been interested in these issues for some time and provided the initial discussions and impetus for this project; the committee hopes that other groups in and especially beyond agriculture colleges will devote the same attention to the reform of undergraduate education in agriculture.

The Role of Universities

This report and its recommendations extend beyond the college of agriculture to the entire university. Agriculture colleges will need to collaborate with other parts of their institutions to offer introductory-level courses that can serve students in a variety of majors, and they should take advantage of opportunities to participate more fully in general education and extend the reach of agriculture. In addition, many of the policies and practices that hamper reform in colleges of agriculture are present throughout the university. The committee hopes that agriculture colleges can lead the way in reforming tenure and promotion practices, implementing active learning, and providing students with greater access to and awareness of career opportunities, but it will be imperative for universities as a whole to address these issues.

It is vital that institutions give these issues high priority. Most institutions will claim that undergraduate education is one of the top priorities, but do their actions demonstrate their commitment? How are decisions made? Where are resources allocated? Which criteria are used to hire and promote faculty, to establish new programs, and to construct new buildings? Institutions will need to back up their spoken commitments and mission statements with action.

The Role of Industry and Other Employers

Although industry has served as an important consumer of agriculture graduates, employers have rarely played a large role in education despite a general concern in industry that today's agriculture graduates do not meet the needs of today's employers. The committee believes that industry and other employers should play a more direct role in the reform of undergraduate education in agriculture. Only by being more involved in education will industry have the opportunity to provide input with respect to the skills and competences that agriculture colleges should be instilling in their students.

The committee has addressed several recommendations to employers and urges companies with an interest in food and agriculture to take a leadership role in discussions, advocacy, and support for high-quality undergraduate education. The committee calls on employers to be a full-fledged partner in the educational processes and to help to implement the changes that are necessary for preparing graduates who have the skills necessary to work in the food and fiber systems, to work across international boundaries in a global marketplace, and to become more educated consumers and more active citizens. Employers can also foster interactions with academic institutions, for example, by offering student internships, supporting career workshops and job-shadowing opportunities, and facilitating exchanges of academic researchers and industry professionals—including sabbatical opportunities and encouragement for food and agriculture professionals to seek visiting-faculty or adjunct-faculty positions.

The Role of Government and Other Funding Agencies

Government and other funding agencies have an obvious influence on agriculture education. For example, the USDA provides critical funding to land-grant universities through the Cooperative State Research, Education, and Extension Service, the U.S. Department of Education supports the National FFA Organization, NSF supports research and programs related to undergraduate education, and a variety of private foundations support education and agriculture. Despite that investment, the committee asked whether additional roles could be played by federal agencies and other funders—roles that could benefit undergraduate education. Although additional resources are often helpful, the committee believes that refocusing small amounts of funds or tweaking the criteria for existing funding programs may produce important rewards with minimal new investment. Moreover, as agriculture has become increasingly science-based, the committee hopes that agriculture will be fully embraced by agencies that support science education in general.

The Role of Professional Societies

As stewards of the discipline, professional societies have an important role to play in speaking on behalf of those in a given field of study. They also play an essential role in bringing together faculty across institutional boundaries and are therefore in a unique position to effect change nationwide. One of the most natural roles for professional societies is to provide discipline-specific information and resources, including maintenance of repositories of relevant teaching materials and sponsorship of workshops targeted to specific fields of study. It will also be important for professional societies to raise the profile of education and of education scholarship in their disciplines.

Professional societies have a number of unique resources that allow the dissemination of ideas through a discipline, including scholarly journals that offer opportunities for dissemination and discussion of new ideas and that allow scholars to read the ideas of others and to publish their own, professional meetings and conferences that bring together hundreds or thousands of practitioners and allow face-to-face meetings and informal conversations that are essential for moving ideas further, and the ability for a discipline to speak with one voice, to support new ideas, and to advocate for positions of common concern. As recommended above, professional societies can give high priority to education and education reform, demonstrating this commitment by giving space to educational topics and papers in their journals and newsletters, offering sessions and prime speaking slots at their meetings and conferences to education topics and speakers, and considering how they can promote and implement education reform nationwide.

The Role of Commodity Groups

Several participants in the Leadership Summit mentioned the importance of state-level agricultural organizations and commodity groups in influencing university decision-making and, in particular, of being barriers to change. The committee believes these groups may be a source of powerful leverage and hopes that they can be encouraged to think broadly about the needs for educating the next generation of professionals in food and agriculture. Engaging those groups in discussions that extend beyond the needs of a single department, crop, or industry can help to provide the consensus needed to move universities forward in a more integrated fashion—the interdisciplinarity called for in Chapter 4.

The Role of Other Stakeholders

Listing employers, professional societies, and commodity groups only scratches the surface of the array of stakeholders with an interest in the issues who should be brought into discussions. The future of agriculture depends on the education and preparation of the next generation of professionals and citizens, and it is essential that all the stakeholders be brought into the conversation. Alumni, donors, boards of trustees, community members, and others all have important roles to play not only in influencing decisions about what should be changed but in helping to bring about that change. All groups should be encouraged to think beyond their individual interests and to focus on the future of the agricultural education enterprise as a whole. If agriculture colleges and disciplines cannot remain vibrant, the future of the entire food and agriculture system is threatened.

CONTINUING THE CONVERSATION

The changes recommended here will not all be achieved immediately; there will need to be a continuing conversation as plans are implemented and the context continues to evolve. The committee hopes that a continuing national conversation will encourage constant sharing of best practices and implementation experiences and will serve as an opportunity for accountability. If academic institutions, food and agricultural industries, professional societies, and others report on their progress periodically, it will not only continue the momentum but provide constant encouragement of action and reform.

The community has already taken steps to continue the conversation of the Leadership Summit. For example, Texas A&M University organized the 2007 National Conference on Changing Higher Education in Agriculture and Related Sciences with the theme "From Dialogue to Action—Reinventing Teaching and Learning."[2] It is hoped that this important follow-up meeting will be the first of many steps, and the committee hopes that the interest shown by both APLU and USDA will continue—supplemented by interest from other national groups that extend beyond land-grant institutions. APLU and USDA have an obvious national and cross-disciplinary interest in the issues, and the committee hopes that regional consortia and professional societies will continue to discuss them in geographic regions and in disciplines. Students, employers, and other stakeholders should be fully engaged in follow-up discussions and specifically invited to participate.

[2]See <http://tti.tamu.edu/conferences/aghe/> for more information about this conference, which was held June 11–13, 2007, in College Station, Texas.

References

AAAS (American Association for the Advancement of Science). 1993. *Benchmarks for Science Literacy*. Project 2061. New York: Oxford University Press. <http://www.project2061.org/publications/bsl/default.htm>.

AAAS. 2004. *Invention and Impact: Building Excellence in Undergraduate Science, Technology, Engineering and Mathematics (STEM) Education*. Washington, D.C.: AAAS. <http://www.aaas.org/publications/books_reports/CCLI/>.

Allen, I. Elaine, and Jeff Seaman. 2007. *Online Nation: Five Years of Growth in Online Learning*. Babson Survey Research Group and The Sloan Consortium. <http://www.sloan-c.org/publications/survey/pdf/online_nation.pdf>.

American Indian College Fund. 1996. Unpublished profiles of tribal colleges.

Armstrong, Norris, Shu-Mei Chang, and Marguerite Brickman. 2007. Cooperative learning in industrial-sized biology classes. *CBE–Life Sciences Education* 6(2): 163-171.

Astin, Alexander W., Lori K. Vogelgesang, Elaine K. Ikeda, and Jennifer A. Yee. 2000. *How Service Learning Affects Students*. Los Angeles: University of California Higher Education Research Institute.

Barber, Maryfran, and David Njus. 2007. Clicker evolution: Seeking intelligent design. *CBE–Life Sciences Education* 6(1): 1-8.

Battistoni, Richard. 2001. *Civic Engagement Across the Curriculum: A Resource Book for Service Learning Faculty in All Disciplines*. Providence, RI: Campus Compact.

Bauer, Karen W., and Joan S. Bennett. 2003. Alumni perceptions used to assess the undergraduate research experience. *Journal of Higher Education* 74(2, March/April): 210-230.

Beatty, Ian D., William J. Leonard, William J. Gerace, and Robert J. Dufresne. 2006. Question driven instruction: Teaching science (well) with an audience response system. Chapter 7 in *Audience Response Systems in Higher Education: Applications and Cases* (David A. Banks, ed.). Hershey, PA: Information Science Publishing.

Beichner, R., L. Bernold, E. Burniston, P. Dail, R. Felder, J. Gastineau, M. Gjertsen, and J. Risley. 1999. Case study of the physics component of an integrated curriculum. *American Journal of Physics, Physics Education Research Supplement* 67(7, July): S16-S24.

Blong, John T., and Honey H. Bedell. 1997. Iowa's community colleges: 32 years of serving the educational needs of Iowans. *Community College Journal of Research and Practice* 21(6): 535-541.

Boland, Mike, and Jay Akridge. 2006. "National Food and Agribusiness Management Education Commission." Presentation to U.S. Department of Agriculture, Washington, D.C., October 24, 2006.

Boyer Commission on Educating Undergraduates in the Research University. 1998. *Reinventing Undergraduate Education: A Blueprint for America's Research Universities*. <http://naples.cc.sunysb.edu/Pres/boyer.nsf/>.

Boyer, Ernest L. 1990. *Scholarship Reconsidered: Priorities of the Professoriate*. San Francisco: Jossey-Bass.

Bruff, Derek. 2009. *Teaching with Classroom Response Systems: Creating Active Learning Environments*. San Francisco: Jossey-Bass.

Bush, S.D., N.J. Pelaez, J.A. Rudd, M.T. Stevens, K.S. Williams, D.E. Allen, and K.D. Tanner. 2006. On hiring science faculty with education specialties for your science (not education) department. *CBE–Life Sciences Education* 5(4): 297-305.

Caldwell, Jane E. 2007. Clickers in the large classroom: Current research and best-practice tips. *CBE–Life Sciences Education* 6(1): 9-20.

Cannon, John G., Thomas W. Broyles, and John H. Hillison. 2006. The role of agriculture in reaching gifted and talented students. *NACTA Journal* 50(3): 2-7.

Capon, Noel, and Deanna Kuhn. 2004. What's so good about problem-based learning? *Cognition and Instruction* 22(1): 61-79.

Case, Linda B. 1999. Transfer opportunity program. Written Testimony [to the] Little Hoover Commission Public Hearing on Community Colleges (ERIC Document ED427824).

Chambers, Sharon M., Sandy R. Weeks, and Larry J. Chaloupka. 2003. Community colleges and higher education: A joint venture. *Community College Journal of Research and Practice* 27(1): 51-55.

Chatman, Steve. 2001. Take the community college route to a selective public university degree: Student affairs research and information (ERIC Document ED457948). Paper presented at the 41st Annual Meeting of the Association for Institutional Research, Long Beach, CA, June 3-6.

Colbert, James T., Joanna K. Olson, and Michael P. Clough. 2007. Using the Web to encourage student-generated questions in large-format introductory biology classes. *CBE–Life Sciences Education* 6(1): 42-48.

Cole, W.M. 2004. Accrediting culture: Legitimacy pressure and resource exigencies at tribal colleges and universities. Unpublished manuscript.

COSEPUP (Committee on Science, Engineering, and Public Policy). 2004. *Facilitating Interdisciplinary Research*. Washington, D.C.: The National Academies Press.

Crofts, Richard A. 1997. History of 2-year postsecondary education in Montana. *Community College Journal of Research and Practice* 21(2): 203-227.

Crouch, Catherine H., and Eric Mazur. 2001. Peer Instruction: Ten years of experience and results. *American Journal of Physics* 69(9, September): 970-977.

DiMaria, Joseph P. 1998. Creating model partnerships that help attract and retain students. Agreements between two-year and four-year colleges. A community college prospective of networking for student success (ERIC Document ED424888).

Dimitri, Carolyn, Anne Effland, and Neilson Conklin. 2005. *The 20th Century Transformation of U.S. Agriculture and Farm Policy*. Economic Information Bulletin No. 3. U.S. Department of Agriculture Economic Research Service. <http://www.ers.usda.gov/Publications/EIB3/>.

Esters, Levon T., and Blannie E. Bowen. 2004. Factors influencing enrollment in an urban agricultural education program. *Journal of Career and Technical Education* 21(1): 25-37.

Fagen, Adam P., Catherine H. Crouch, and Eric Mazur. 2002. Peer Instruction: Results from a range of classrooms. *The Physics Teacher* 40(4): 206-209.

Fies, Carmen, and Jill Marshall. 2006. Classroom response systems: A review of the literature. *Journal of Science Education and Technology* 15(1): 101-109.

Friedman, Thomas L. 2005. *The World is Flat: A Brief History of the Twenty-First Century*. New York: Farrar, Straus and Giroux.

Gelmon, Sherril B., Barbara A. Holland, Amy Driscoll, Amy Spring, and Seanna Kerrigan. 2001. *Assessing Service-Learning and Civic Engagement: Principles and Techniques*. Providence, RI: Campus Compact.

Gijbels, David, Filip Dochy, Piet Van den Bossche, and Mien Segers. 2005. Effects of problem-based learning: A meta-analysis from the angle of assessment. *Review of Educational Research* 75(1): 27-61.

Gilmore, Jeffrey L., Allan D. Goecker, Ella Smith, and P. Gregory Smith. 2006. *Shifts in the Production and Employment of Baccalaureate Degree Graduates from U.S. Colleges of Agriculture and Natural Resources, 1990-2005*. Background paper for "A Leadership Summit to Effect Change in Teaching and Learning," Washington, D.C., October 3-5, 2006. <http://dels.nas.edu/banr/summit/GilmorePaper.pdf>.

Goecker, Allan D., Jeffrey L. Gilmore, Ella Smith, and P. Gregory Smith. 2005. *Employment Opportunities for College Graduates in the U.S. Food, Agricultural and Natural Resources System, 2005-2010*. U.S. Department of Agriculture. <http://faeis.ahnrit.vt.edu/supplydemand/2005-2010/>.

Gonzalez, Jorge A. 2006. "Agricultural Programs: Are They Able to Adapt for the Future?" CSREES Faculty Fellow presentation at USDA. Washington, D.C., August 2006.

Graves, Fritz. 1998. The Midwest Poultry Consortium. *Poultry Science* 77: 211-213.

Green, Madeleine, Dao T. Luu, and Beth Burris. 2008. *Mapping Internationalization on US Campuses: 2008 Edition*. Washington, D.C.: American Council on Education.

Guess, Andy. 2008. "Extension Goes National—and Online." *Inside Higher Ed*, June 9, 2008. <http://insidehighered.com/news/2008/06/09/extension>.

Hake, Richard R. 1998. Interactive-engagement versus traditional methods: A six-thousand-student survey of mechanics test data for introductory physics courses. *American Journal of Physics* 66(1, January): 64-74.

Handelsman, Jo, Diane Ebert-May, Robert Beichner, Peter Bruns, Amy Chang, Robert DeHaan, Jim Gentile, Sarah Lauffer, James Stewart, Shirley M. Tilghman, and William B. Wood. 2004. Scientific Teaching. *Science* 304(5670): 521-522.

Henderson, Charles, and Melissa H. Dancy. 2007. Barriers to the use of research-based instructional strategies: The influence of both individual and situational characteristics. *Physical Review Special Topics—Physics Education Research* 3(2): 020102.

Henderson, Charles, and Melissa H. Dancy. 2008. Physics faculty and educational researchers: Divergent expectations as barriers to the diffusion of innovations. *American Journal of Physics* 76(1): 79-91.

Hunter, Anne-Barrie., Sandra L. Laursen, and Elaine Seymour. 2007. Becoming a scientist: The role of undergraduate research in students' cognitive, personal, and professional development. *Science Education* 91: 36-74.

Ignash, Jan M., and Barbara K. Townsend. 2000. Evaluating state-level articulation agreements according to good practice. *Community College Review* 28(3): 1-21.

James, Donna Walker, Sonia Jurich, and Steve Estes. 2001. *Raising Minority Academic Achievement: A Compendium of Education Programs and Practices*. Washington, D.C.: American Youth Policy Forum. <http://www.aypf.org/publications/rmaa/index.html>.

Jaschik, Scott. 2008. "The Truly Interdisciplinary Search." *Inside Higher Ed*, September 5. <http://www.insidehighered.com/news/2008/09/05/michtech>.

Johnson, David W., and Roger T. Johnson. 1989. *Cooperation and Competition: Theory and Research*. Edina, MN: Interaction Book Company.

Johnson, David W., Roger T. Johnson, and Karl A. Smith. 1991. *Active Learning: Cooperation in the College Classroom*. Edina, MN: Interaction Book Company.

Jordan, Nicholas R., David A. Andow, and Kristin L. Mercer. 2005. New concepts in agroecology: A service-learning course. *Journal of Natural Resources and Life Science Education* 34: 83-92.

Kardash, CarolAnne M. 2000. Evaluation of undergraduate research experience: Perceptions of undergraduate interns and their faculty mentors. *Journal of Educational Psychology* 92(1): 191-201.

Kirschner, Paul A., John Sweller, and Richard E. Clark. 2006. Why minimal guidance during instruction does not work: An analysis of the failure of constructivist, discovery, problem-based, experiential, and inquiry-based teaching. *Educational Psychologist* 41(2): 75-86.

Kisker, Carrie B. 2007. Creating and sustaining community college–university transfer partnerships. *Community College Review* 34(4): 282-301.

Knight, Jennifer K., and William B. Wood. 2005. Teaching more by lecturing less. *Cell Biology Education* 4(4): 298-310.

Laufgraben, Jodi Levine, and Nancy S. Shapiro. 2004. *Sustaining and Improving Learning Communities*. San Francisco: Jossey-Bass.

Lopatto, David. 2003. The essential features of undergraduate research. *Council on Undergraduate Research Quarterly* 24: 139-142.

Lopatto, David. 2004. Survey of Undergraduate Research Experiences (SURE): First findings. *Cell Biology Education* 3(4): 270-277.

Lopatto, David. 2007. Undergraduate research experiences support science career decisions and acive learning. *CBE–Life Sciences Education* 6(4): 297-306.

Mazur, Eric. 1997. *Peer Instruction: A User's Manual*. Upper Saddle River, NJ: Prentice Hall.

McNeal, Ann P., and Charlene D'Avanzo, eds. 1997. *Student-Active Science: Models of Innovation in College Science Teaching*. Proceedings on the NSF sponsored conference on Inquiry Approaches to Science Teaching (Hampshire College, June 1996). Fort Worth, TX: Saunders College Publishing.

McNeill, Joyce H., and Pamela K. Payne. 1996. Cooperative learning groups at the college level: Applicable learning. Paper presented at the Division for Early Childhood, International Early Childhood Conference on Children with Special Needs (Phoenix, AZ, December 1996). ERIC Document #ED404920.

Michael, Joel. 2006. Where's the evidence that active learning works? *Advances in Physiology Education* 30: 159-167.

Michaelsen, Larry K., Arletta Bauman Knight, and L. Dee Fink., eds. 2002. *Team-Based Learning: A Transformative Use of Small Groups*. Westport, CT: Praeger Publishers.

Michigan Technological University. 2008. "Michigan Tech Names 3 Endowed Chairs, Adds Sustainability Faculty." Media Relations Story #719. <http://www.admin.mtu.edu/urel/news/media_relations/719/>.

Morrill, Justin S. 1887. Address. In *Addresses delivered at the Massachusetts Agricultural College, June 21st, 1887, on the 25th Anniversary of the Passage of the Morrill Land Grant Act*. Amherst, MA: J.E. Williams, Book and Job Printer.

Novak, Gregor M., Evelyn T. Patterson, Andrew D. Garvin, and Wolfgang Christian. 1999. *Just-in-Time Teaching: Blending Active Learning with Web Technology*. Upper Saddle River, NJ: Prentice Hall.

NRC (National Research Council). 1992. *Agriculture and the Undergraduate*. Washington, D.C.: National Academy Press. <http://www.nap.edu/catalog/1986.html>.

NRC. 1996a. *From Analysis to Action: Undergraduate Education in Science, Mathematics, Engineering, and Technology: Report of a Convocation*. Washington, D.C.: National Academy Press. <http://www.nap.edu/catalog/9128.html>.

NRC. 1996b. *National Science Education Standards*. Washington, D.C.: National Academy Press. <http://www.nap.edu/catalog/4962.html>.

NRC. 1997. *Science Teaching Reconsidered: A Handbook*. Washington, D.C.: National Academy Press. <http://www.nap.edu/catalog/5287.html>.

NRC. 1999a. *How People Learn: Brain, Mind, Experience, and School*. John D. Bransford, Ann L. Brown, and Rodney R. Cocking, eds. Washington, D.C.: National Academy Press. <http://www.nap.edu/catalog/6160.html>.

NRC. 1999b. *How People Learn: Bridging Research and Practice*. M. Suzanne Donovan, John D. Bransford, and James W. Pellegrino, eds. Washington, D.C.: National Academy Press. <http://www.nap.edu/catalog/9457.html>.

NRC. 1999c. *Transforming Undergraduate Education in Science, Mathematics, Engineering, and Technology*. Washington, D.C.: National Academy Press. <http://www.nap.edu/catalog/6453.html>.

NRC. 2003a. *Bio2010: Transforming Undergraduate Education for Future Research Biologists*. Washington, D.C.: The National Academies Press. <http://www.nap.edu/catalog/10497.html>.

NRC. 2003b. *Evaluating and Improving Undergraduate Teaching in Science, Technology, Engineering, and Mathematics*. Washington, D.C.: The National Academies Press. <http://www.nap.edu/catalog/10024.html>.

NRC. 2003c. *Improving Undergraduate Instruction in Science, Technology, Engineering, and Mathematics: Report of a Workshop*. Richard A. McCray, Robert L. DeHaan, and Julie Anne Schuck, eds. Washington, D.C.: The National Academies Press. <http://www.nap.edu/catalog/10711.html>.

NRC. 2005a. *America's Lab Report: Investigations in High School Science*. Susan R. Singer, Margaret L. Hilton, and Heidi A. Schweingruber, eds. Washington, D.C.: The National Academies Press. <http://ww.nap.edu/catalog/11311.html>.

NRC. 2005b. *How People Learn: Science in the Classroom*. M. Suzanne Donovan and John D. Bransford, eds. Washington, D.C.: The National Academies Press. <http://www.nap.edu/catalog/11102.html>.

NSB (National Science Board). 2007. *A National Action Plan for Addressing the Critical Needs of the U.S. Science, Technology, Engineering, and Mathematics Education System*. NSB-07-114. Arlington, VA: National Science Foundation. <http://www.nsf.gov/nsb/documents/2007/stem_action.pdf>.

Pfund, Christine, Sarah Miller, Kerry Brenner, Peter Bruns, Amy Chang, Diane Ebert-May, Adam P. Fagen, Jim Gentile, Sandra Gossens, Ishrat M. Khan, Jay B. Labov, Christine Maidl Pribbenow, Millard Susman, Lillian Tong, Robin Wright, Robert T. Yuan, William B. Wood, and Jo Handelsman. 2009. Summer Institute to improve university science teaching. *Science* 324(5296): 470-471.

Preszler, Ralph W., Angus Dawe, Charles B. Shuster, and Michèle Shuster. 2007. Assessment of the effects of student response systems on student learning and attitudes over a broad range of biology courses. *CBE–Life Sciences Education* 6(1): 29-41.

Rifkin, Tronie. 2000. Improving articulation policy to increase transfer. Education Commission of the States policy paper. <http://www.eric.ed.gov/ERICDocs/data/ericdocs2sql/content_storage_01/0000019b/80/16/24/cf.pdf>.

Rueckert, Linda. 2002. Report from CUR 2002: Assessment of Research. *Council on Undergraduate Research Quarterly* 23: 10-11.

Russell, Susan H., Mary P. Hancock, and James McCullough. 2007. Benefits of undergraduate research experiences. *Science* 316(5824): 548-549.

Seymour, Elaine, and Nancy M. Hewitt. 2000. *Talking about Leaving: Why Undergraduates Leave the Sciences*. Boulder, CO: Westview Press.

Shapiro, Nancy S., and Jodi H. Levine. 1999. *Creating Learning Communities: A Practical Guide to Winning Support, Organizing for Change, and Implementing Change*. San Francisco: Jossey-Bass.

Singer, Susan R. 2002. Learning and teaching centers: Hubs of educational reform. Chapter 10 (pp. 59-64) in *Building Robust Learning Environments in Undergraduate Science, Technology, Engineering, and Mathematics: New Directions for Higher Education, No. 119* (Jeanne L. Narum and Kate Conover, eds.). San Francisco: Jossey-Bass.

Smith, Barbara Leigh, Jean MacGregor, Roberta Matthews, and Faith Gabelnick. 2004. *Learning Communities: Reforming Undergraduate Education*. San Francisco: Jossey-Bass.

Snyder, Thomas D., Sally A. Dillow, and Charlene M. Hoffman. 2009. *Digest of Education Statistics 2008* (NCES 2009-020). National Center for Education Statistics, Institute of Education Sciences, U.S. Department of Education, Washington, D.C.

Taylor, Kathe, William S. Moore, Jean MacGregor, and Jerri Lindblad, eds. 2003. *Learning Community Research and Assessment: What We Know Now*. Olympia, WA: The Evergreen State College, Washington Center for Improving the Quality of Undergraduate Education, in cooperation with the American Association for Higher Education.

Taylor, Mark. 2008. Meet the students: Finding common ground between student and institutional goals. Pages 3-9 in *Finding Common Ground: Programs, Strategies, and Structures to Support Student Success, Vol. 3*. Chicago: The Higher Learning Commission.

Tobias, Sheila. 1992. *Revitalizing Undergraduate Science: Why Some Things Work and Most Don't*. Tucson, AZ: Research Corporation.

Tobias, Sheila. 1994. *They're Not Dumb, They're Different: Stalking the Second Tier*. Tucson, AZ: Research Corporation.

Townsend, Barbara K., and Jan M. Ignash, eds. 2003. *The Role of the Community College in Teacher Education*. New Directions for Community Colleges, No. 121. San Francisco: Jossey-Bass.

U.S. Department of Education. 2006. *A Test of Leadership: Charting the Future of U.S. Higher Education*. Washington, D.C. <http://www.ed.gov/about/bdscomm/list/hiedfuture/index.html>.

Wood, William, and James Gentile. 2003. Meeting report: The first National Academies Summer Institute for Undergraduate Education in Biology. *Cell Biology Education* 2(4): 207-209.

Wood, William B., and Jo Handelsman. 2004. Meeting report: The 2004 National Academies Summer Institute on Undergraduate Education in Biology. *Cell Biology Education* 3(4): 215-217.

Wright, John C., Susan B. Millar, Steve A. Koscuik, Debra L. Penberthy, Paul H. Williams, and Bruce E. Wampold. 1998. A novel strategy for assessing the effects of curriculum reform on student competence. *Journal of Chemical Education* 75(8): 986-992.

Zirkle, Chris, Jennifer Brenning, and John Marr. 2006. An Ohio model for articulated teacher education. *Community College Journal of Research and Practice* 30(5/6): 501-512.

Appendixes

Appendix A

Statement of Task

The National Academies will conduct a study, including a major two-day summit of educators, employers, and others, to explore opportunities for institutions of higher education to improve the learning experience of undergraduate students pursuing careers at the intersection of agriculture, environmental and life sciences, and their related disciplines. The summit will examine innovations in teaching, learning, and the curriculum that are adaptive to differences in student backgrounds, attitudes, and expectations, and that better equip graduates with knowledge and skills appropriate for multiple career paths and demands. Following the summit, a committee of the National Academies will prepare a report that identifies opportunities to effect change in undergraduate programs that will enable those programs to produce a flexible, well-prepared workforce that is appropriately skilled, socially responsive, and technically proficient.

Appendix B

Leadership Summit Information[1]

A LEADERSHIP SUMMIT TO EFFECT CHANGE IN TEACHING AND LEARNING
Board on Agriculture and Natural Resources and Board on Life Sciences

October 3-5, 2006
National Academy of Sciences Building
2100 C Street, NW
Washington, D.C.

Agenda

All plenary sessions will be held in the Auditorium.
Breakout sessions will be throughout the building.

TUESDAY, OCTOBER 3, 2006

12:00–2:00 p.m. Summit Registration in the C Street lobby

2:00 p.m. **Welcome and introduction**

- Ralph J. Cicerone, *President, National Academy of Sciences*
- James L. Oblinger, *Chancellor, North Carolina State University (Committee Chair)*

[1]Additional information about the Leadership Summit, including speaker presentations, poster abstracts, and a participant list, is available at <http://www.nationalacademies.org/summit>.

2:25 p.m. **The USDA interest in agriculture education**

Session Chair: W.R. "Reg" Gomes, *Vice President for Agriculture and Natural Resources, University of California System; Chair, Board on Agriculture and Natural Resources*

- Gale A. Buchanan, *Under Secretary for Research, Education, and Economics, U.S. Department of Agriculture*

2:50 p.m. **A call to action: Opening keynotes from the perspectives of industry and academia**

Session Chair: Susan J. Crockett, *Vice President and Senior Technology Officer, Health and Nutrition, General Mills, Inc. (committee member)*

A Look Ahead

- Gary Rodkin, *Chief Executive Officer, ConAgra Foods, Inc.*

University of the Future

- Michael V. Martin, *President, New Mexico State University (committee member)* [standing in for Peter McPherson, *President, National Association of State Universities and Land-Grant Colleges (NASULGC)*]

3:50 p.m. **Undergraduate education: Reflections on the past and future**

Session Chair: Vernon B. Cardwell, *Morse-Alumni Distinguished Teaching Professor, Department of Agronomy and Plant Genetics, University of Minnesota (committee member)*

- C. Eugene Allen, *Distinguished Teaching Professor and Former Dean of the College of Agriculture; Former Vice President for Agriculture, Forestry, and Home Economics; and Former Provost for Professional Studies, University of Minnesota*

4:20 p.m. **Introduction to breakout session 1**

- A. Charles Fischer, *Past President and Chief Executive Officer, Dow AgroSciences LLC (committee member)*

4:30 p.m. **Breakout session 1: Defining the goals of an education in agriculture**

This breakout session will seek to answer questions such as: What are the goals of an undergraduate education in agriculture? How does it differ from other science degrees? What are we preparing students for?

6:00–7:30 p.m. **Welcome reception**

Poster session in the Upstairs Gallery (2nd floor) and Great Hall Foyer

WEDNESDAY, OCTOBER 4, 2006

8:00 a.m. Continental Breakfast in the Great Hall

8:30 a.m. **Overview of program for the day**

- James L. Oblinger, *Chancellor, North Carolina State University (committee chair)*

8:45 a.m. **Remarks from U.S. Secretary of Agriculture**

- The Honorable Mike Johanns, *Secretary of Agriculture*

9:15 a.m. **Panel on the agriculture classroom**

This panel will include current discussions on teaching and learning.

Session Chair: Janet Guyden, *Associate Vice President of Research and Dean of Graduate Studies, Grambling State University (committee member)*

How people learn

- M. Suzanne Donovan, *Program Director, Strategic Education Research Partnership Institute; Study Director,* How People Learn, *National Research Council*

Re-envisioning our classrooms as learning laboratories

- Robin Wright, *Professor of Genetics, Cell Biology and Development; Associate Dean for Faculty and Academic Affairs, College of Biological Sciences, University of Minnesota*

Moving towards institutional change in teaching and learning

- Jose P. Mestre, *Professor of Physics and Educational Psychology, University of Illinois at Urbana-Champaign*

The convergence of culture and pedagogy: Implications for teaching and learning in STEM disciplines

- Wynetta Y. Lee, *Associate Vice President for Academic Planning, Research & Graduate Studies, California State University, Monterey Bay*

10:45 a.m. Break

11:05 a.m. **Agriculture education in the context of other disciplines**

This session will feature speakers from primarily non-agriculture disciplines on the ways that agriculture contributes to research and development in other fields, highlighting the need to work together on reforming undergraduate education for mutual benefit.
Session Chair: Michael W. Hamm, *C.S. Mott Professor of Sustainable Agriculture, Michigan State University (committee member)*

Rural Development
- John C. Allen, *Director, Western Rural Development Center; Professor of Sociology, Social Work, and Anthropology, Utah State University* [by videoconference]

Regulatory Affairs: Two views
- Jay Ellenberger, *Associate Director, Field and External Affairs Division, Office of Pesticide Programs, U.S. Environmental Protection Agency*
- Sally L. Shaver, *Associate Counselor for Agricultural Policy, Office of Air and Radiation, U.S. Environmental Protection Agency*

Medicine
- Jay Moskowitz, *Associate Vice President for Health Sciences Research; Vice Dean for Research and Graduate Studies, College of Medicine, The Pennsylvania State University*

Nutrition
- Marion Nestle, *Paulette Goddard Professor of Nutrition, Food Studies, and Public Health, New York University*

12:30 p.m. **Introduction to breakout sessions 2 and 3**
- Karen Gayton Swisher, *President, Haskell Indian Nations University (committee member)*

12:40 p.m. Box lunches available in the Great Hall

1:00 p.m. **Breakout session 2: Overcoming barriers to interdisciplinary** (during lunch)
For these breakout sessions, participants will stay in their institutional teams and work with other teams to identify barriers to working across disciplines and opportunities to overcome those barriers.

2:30 p.m. Break

3:00 p.m.

Breakout session 3: Concurrent topics, best practices, and implementation

These concurrent sessions will feature presentations on sample programs and best practices to address particular needs and objectives followed by a discussion on the opportunities and challenges to implementing similar objectives at other institutions.

- Academic-industry partnerships domestically and abroad through internships and cooperative education
  - Thomas M. Akins, *Executive Director, Division of Professional Practice, Georgia Institute of Technology*
- Academic-industry partnerships: UC-Davis Program in Viticulture & Enology
  - Andrew L. Waterhouse, *John E. Kinsella Chair in Food, Nutrition and Health and Interim Chair, Department of Viticulture & Enology, University of California, Davis*
- Articulation between community colleges and four-year institutions
  - Jerry Bolton, *Dean of Agriculture, Kirkwood Community College, Cedar Rapids, IA*
- Faculty development: Project Kaleidoscope
  - Jeanne Narum, *Director, Project Kaleidoscope*
- Globalization of the science classroom
  - Robert T. Yuan, *Professor Emeritus of Cell Biology & Molecular Genetics, University of Maryland, College Park*
  - Vanessa Sitler, *senior undergraduate student in business management, Robert H. Smith School of Business, University of Maryland, College Park*
- How can we value teaching at tenure time?
  - Caitilyn Allen, *Professor of Plant Pathology, University of Wisconsin–Madison*

- International experiences outside of the United States
  - Frank Fear, *Senior Associate Dean, College of Agriculture and Natural Resources, Michigan State University*
  - Paul Roberts, *Director of Study Abroad and International Training, College of Agriculture and Natural Resources, Michigan State University*
- Partnerships for sustainable development of agriculture: Green Lands, Blue Waters Initiative
  - Nicholas R. Jordan, *Professor of Agroecology, Department of Agronomy & Plant Genetics, University of Minnesota*
- Professional Science Masters
  - Paul D. Tate, *Senior Scholar in Residence and Co-director of the Professional Science Master's Initiative, Council of Graduate Schools*

4:30 p.m. **Reporting back on breakout session 2**

Groups report back on their lunchtime discussion.

- Moderator: Susan Singer, *Laurence McKinley Gould Professor of the Natural Sciences, Carleton College (committee member)*

5:20 p.m. **Introduction to breakout session 4 (Thursday morning)**

- Levon T. Esters, *Assistant Professor of Agricultural Education and Studies, Iowa State University (committee member)*

5:30 p.m. Adjourn for the day

THURSDAY, OCTOBER 5, 2006

8:00 a.m. Continental Breakfast in the Great Hall

8:30 a.m.	**Breakout session 4: Identifying action items and next steps** These breakout sessions will enable participants to discuss opportunities and responsibilities for implementing change. Participants will divide into common stakeholder groups (e.g., academic administrators, teaching faculty, industry representatives, professional societies). Each group will seek to identify action items, challenges, and needed resources for moving forward with implementation.
10:15 a.m.	Break
10:40 a.m.	**Reporting back on breakout session 4** • Moderator: Patricia Verduin, *Senior Vice President and Director of Product Quality and Development, ConAgra Foods, Inc. (committee member)*
11:30 a.m.	**Summary and wrap-up** Michael V. Martin, *President, New Mexico State University (committee member)* James L. Oblinger, *Chancellor, North Carolina State University (committee chair)*
12:00 p.m.	Adjourn: Thank you for your participation.

The National Academy of Sciences (NAS) building is located along the National Mall in Washington, D.C., close to the Lincoln Memorial and Vietnam Veterans Memorial. The entrance to the building is at 2100 C Street, NW, between 21st and 22nd Streets. Please be aware that C Street is closed to automobile traffic between 21st Street and 23rd Street (NAS is located across from the State Department). Be prepared to show a photo ID to enter the building.

Sponsors for this project: U.S. Department of Agriculture, W.K. Kellogg Foundation, National Science Foundation, Farm Foundation, and American Farm Bureau Foundation for Agriculture

SPEAKER BIOGRAPHIES[2]

Thomas M. Akins is the Executive Director of the Division of Professional Practice at Georgia Tech. He oversees the operation of the nation's largest totally optional cooperative education program as well as the undergraduate professional internship (UPI) program, the Graduate Co-op Program, and the Work Abroad Program. In cooperative education, Tom has made presentations and conducted workshops on the state, regional, national, and international level. He was elected multiple terms to the Faculty Assembly, Academic Senate, and Executive Board (currently the Vice-Chair). Mr. Akins holds memberships in the American Society for Engineering Education (ASEE), the World Association for Cooperative Education, the Cooperative Education and Internship Association, and the National Association of Multicultural Engineering Program Advocates. He has served as Secretary-Treasurer, Chair-Elect, and Chairman of the Cooperative Education Division of ASEE. He is a founding member of the national co-op accrediting body, the Accreditation Council for Cooperative Education (currently serving as President). Mr. Akins is the recipient of the 1998 Borman Award for outstanding service to the field of Cooperative Education, and the 2003 Clement J. Freund Award from ASEE for outstanding contributions to the aims and ideals of cooperative education. Mr. Akins received his MBA from Georgia State University and his Bachelor of Industrial Engineering degree from Georgia Tech.

Caitilyn Allen is Professor of Plant Pathology at the University of Wisconsin–Madison, where she has taught since 1992. She just completed three years on the university-level tenure committee, serving as its chair in 2005–2006. Professor Allen's research lab studies mechanisms of virulence in bacterial pathogens of plants, and she also has an applied research project to develop disease-resistant tomatoes for Central American farmers. She has taught courses on molecular plant–microbe interactions, plant-associated bacteria, and tropical plant pathology, as well as two biology courses for nonscience majors. She received UW–Madison's Distinguished Teaching Award and the American Phytopathological Society's National Award for Excellence in Teaching. Professor Allen was the founding Director of UW's Women in Science and Engineering Residential Program and also holds an appointment in the Women's Studies Program.

[2]Biographies are current as of the time of the Leadership Summit.

C. Eugene Allen is a Distinguished Teaching Professor and Former Dean, Vice President, and Provost of the University of Minnesota. In a distinguished career, he taught more than 3,000 students in his undergraduate and graduate courses, had an internationally recognized research program on the growth of muscle and adipose tissue and their use as meat, and extended these results through his outreach efforts. This resulted in about a hundred scientific publications and more than 40 outreach publications. He has been an invited speaker for hundreds of audiences in different states and countries, has served on numerous program or award review teams, boards of directors, and he has diverse work experiences in 22 countries that are primarily in the developing world. Gene is the recipient of three University of Minnesota teaching awards, two national awards for his research, numerous state and national awards for service, and is an elected Fellow of the American Association for the Advancement of Science and the Institute of Food Technology. Since 1984, he has provided visionary leadership in roles such as dean (1984–1988), director of the agricultural experiment station (1988–1997), vice president (1988–1995), provost (1995–1997), and most recently as associate vice president for international programs (1998–2006). He has frequently given leadership to national and international initiatives or organizations. The most recent examples include national initiatives to internationalize campuses and expand study abroad enrollments. In the 1980s, Gene and three colleagues took the initiative that led to formation of the National Academies' Board on Agriculture (BOA). Later he served for six years on the BOA Board of Directors, plus five National Research Council committees (including the steering committee for the 1992 *Agriculture and the Undergraduate* effort), and he has been an invited speaker for three NAS workshops. In 1989, he was honored as a "Distinguished Centennial Alumni" of the University of Idaho. Dr. Allen is a native of Idaho and received a B.S. degree from the University of Idaho (1961), and M.S. (1963) and Ph.D. (1965) degrees from the University of Wisconsin.

John C. Allen is the Director of the Western Rural Development Center (WRDC) and Professor in the Department of Sociology, Social Work, and Anthropology at Utah State University in Logan. Dr. Allen grew up on a ranch in eastern Oregon. Since that time, he has worked as a farmer and rancher, journalist, market researcher, and professor. Before accepting the position of WRDC Director, he was Director of the Center for Applied Rural Innovation at the University of Nebraska–Lincoln. Dr. Allen's professional activities focus on rural community development, entrepreneurial communities, and natural resource management throughout the West. His research interests include the impact of information age technology on economic development, how communities respond to change, the impact of sustainable agriculture on rural communities, and the role natural resources play in rural development. His research has been adapted to cooperative extension educational programs including *Navigating the Net, Master Navigator, Working More Effectively in Rural Communities, Community Conflict Management,* the *EDGE (Enhancing Developing and Growing Entrepreneurs), Nebraska Annual Rural Poll, Tilling the Soil of Opportunity,* and *Asset Based Community Development.* Dr. Allen received his Ph.D. in sociology from Washington State University, Pullman, M.S. in urban sociology from Portland State University, and B.S. in sociology from Southern Oregon State University.

Jerry Bolton is the dean of Agriculture and Natural Resources at Kirkwood Community College in Cedar Rapids, Iowa. Kirkwood Community College's agriculture department has an enrollment of nearly 800 students in 15 different programs with 25 full-time instructors and a 500-acre teaching lab and contributes to the second highest national ranking in conferring two-year associate agriculture degrees. He has implemented increased math and science skills into agriculture curriculums, worked for increased articulation agreements with university agriculture programs, and successfully procured grants from local, state, and national entities, with the largest being a seven-year $6 million grant from the National Science Foundation for the development of an advanced technology curriculum for agriculture focusing on associate degree colleges. Prior to his position at Kirkwood, he was chair of the department of agriculture and natural resources at Hawkeye Community College in Waterloo, Iowa; a grain elevator, feed, and fertilizer business manager; and a high school vocation agricultural teacher. Mr. Bolton received his M.S. and B.S. from Iowa State University.

Gale A. Buchanan is the Under Secretary for Research, Education, and Economics at the U.S. Department of Agriculture (USDA). Before joining USDA, from March 1995 until his May 2006 confirmation for this position by the U.S. Senate, Dr. Buchanan served as Dean and Director of the College of Agricultural and Environmental Sciences at the University of Georgia. He was Interim Director of the Georgia Agricultural Experiment Stations from 1994 to 1995. Previously he had served as their Associate Director as well as the Resident Director of the Coastal Plain Experiment Station—all affiliated with the University of Georgia—from 1986 to 1994. He was the Dean and Director of the Alabama Agricultural Experiment Station at Auburn

University from 1980 to 1985. He began his full-time academic career in 1965 at Auburn University's Department of Agronomy and Soils, with primary teaching and research responsibilities in weed science. Dr. Buchanan received his Ph.D. in plant physiology from Iowa State University, and his M.S. and B.S. in agronomy from the University of Florida.

Ralph J. Cicerone, president of the National Academy of Sciences, is an atmospheric scientist whose research in atmospheric chemistry and climate change has involved him in shaping science and environmental policy at the highest levels nationally and internationally. His research was recognized on the citation for the 1995 Nobel Prize in chemistry awarded to University of California, Irvine colleague F. Sherwood Rowland. The Franklin Institute recognized his fundamental contributions to the understanding of greenhouse gases and ozone depletion by selecting Cicerone as the 1999 laureate for the Bower Award and Prize for Achievement in Science. One of the most prestigious American awards in science, the Bower also recognized his public policy leadership in protecting the global environment. In 2001, he led a National Academy of Sciences study of the current state of climate change and its impact on the environment and human health, requested by President Bush. The American Geophysical Union awarded him its 2002 Roger Revelle Medal for outstanding research contributions to the understanding of Earth's atmospheric processes, biogeochemical cycles, or other key elements of the climate system. In 2004, the World Cultural Council honored him with another of the scientific community's most distinguished awards, the Albert Einstein World Award in Science.

During his early career at the University of Michigan, Cicerone was a research scientist and held faculty positions in electrical and computer engineering. In 1978 he joined the Scripps Institution of Oceanography at the University of California, San Diego, as a research chemist. From 1980 to 1989, he was a senior scientist and director of the atmospheric chemistry division at the National Center for Atmospheric Research in Boulder, Colorado. In 1989 he was appointed the Daniel G. Aldrich Professor of Earth System Science at the University of California, Irvine, and chaired the department of earth system science from 1989 to 1994. While serving as dean of physical sciences for the next four years, he brought outstanding faculty to the school and strengthened its curriculum and outreach programs.

Prior to his election as Academy president, Cicerone was the chancellor of the University of California, Irvine, from 1998 to 2005. Cicerone is a member of the National Academy of Sciences, the American Academy of Arts and Sciences, and the American Philosophical Society. He served as president of the American Geophysical Union, the world's largest society of earth scientists, and he received its James B. Macelwane Award in 1979 for outstanding contributions to geophysics. He has published about 100 refereed papers and 200 conference papers, and has presented invited testimony to the U.S. Senate and House of Representatives on a number of occasions. Cicerone received his bachelor's degree in electrical engineering from the Massachusetts Institute of Technology where he was a varsity baseball player. Both his master's and doctoral degrees are from the University of Illinois in electrical engineering, with a minor in physics.

M. Suzanne Donovan is Executive Director of the Strategic Education Research Partnership (SERP) Institute, an independent nonprofit organization that began functioning independently of the National Academies in December of 2004. As Associate Director then Director of the National Academies' SERP project, she was co-editor of the project's two reports: *Strategic Education Research Partnership* and *Learning and Instruction: A SERP Research Agenda*. She served as study director and editor of *How Students Learn: History, Math, and Science in the Classroom,* as well as a previous study in the series entitled *How People Learn: Bridging Research and Practice*. She was the study director and co-editor for the NRC report *Minority Students in Special and Gifted Education,* and was a co-editor of *Eager to Learn: Educating our Preschoolers*. Dr. Donovan was previously on the faculty of Columbia University. She has a Ph.D. in public policy from the University of California at Berkeley.

Jay Ellenberger is the associate director of the Field and External Affairs Division at the U.S. Environmental Protection Agency's (EPA's) Office of Pesticide Programs. He has served for more than 25 years with EPA's Office of Pesticide Programs in the areas of regulatory activities, national and international policy and program development, legislation and communications, and homeland security. He leads the agency's homeland security efforts in protecting the food and agriculture sectors and representing the EPA in initiatives to protect these sectors. Mr. Ellenberger holds undergraduate and graduate degrees in animal science and entomology from Pennsylvania State University.

Frank Fear is Senior Associate Dean in the College of Agriculture and Natural Resources at Michigan State University (MSU). Dr. Fear provides oversight to the College's General Fund budget; is responsible for facilitating connections between the College's academic units and the Dean's Office, including working with the units on strategic planning and organization development efforts; and is point person for the College's global programs. He established his academic credentials in the fields of community and organization development. He served as chairperson of the Department of Resource Development—an academic department devoted to community and natural resource development. He worked in the Office of the Vice Provost for University Outreach, helping to develop an intellectual foundation and strategic plan for MSU's outreach efforts—an approach that informs the University's work to this day. Dr. Frank served as acting associate director of MSU Extension, and was the inaugural chairperson of The Liberty Hyde Bailey Scholars Program—a distinctive, college-wide undergraduate program. The John Templeton Foundation and Phi Kappa Phi have recognized the program for the way it promotes undergraduate student and faculty development through collaborative learning. Dr. Fear is the lead author of the recently published book *Coming to Critical Engagement*. In 2006, Frank was named a Senior Fellow in Outreach and Engagement at MSU. He has also been involved in organizational consulting and civic affairs, notably as a consultant with the W.K. Kellogg Foundation and as president and chief executive officer of the Greater Lansing Food Bank (2004–2006). Dr. Fear received his Ph.D. in sociology from Iowa State University.

The Honorable Mike Johanns was sworn in as the 28th Secretary of the U.S. Department of Agriculture (USDA) on January 21, 2005. Secretary Johanns' strong agricultural roots stretch back to his childhood. He was born in Iowa and grew up doing chores on his family's dairy farm. As the son of a dairy farmer, he developed a deep respect for the land and the people who work it. He still describes himself as "a farmer's son with an intense passion for agriculture." That passion has been evident during Johanns' tenure as Secretary of Agriculture. Days after he took office, he began working with U.S. trading partners to reopen their markets to U.S. beef. Nearly 119 countries had closed their markets after a single finding of a BSE-infected cow in the United States in 2003. Within his first year, Johanns convinced nearly half that number to reopen markets.

To improve access to markets, he has traveled the world, participating in World Trade Organization negotiations and promoting the successful passage of the Dominican Republic-Central American Free Trade Agreement. To fight obesity he launched the interactive, bilingual MyPyramid.com, a motivational and interactive food guidance system. A companion site for children is also available. To aid producers he has led the effort to provide timely assistance after the devastating hurricane season of 2005.

He has promoted the use and promise of renewable fuels and he has supported conservation by expanding USDA's conservation commitment. He has also worked to educate and prepare the country for the potential onset of avian flu.

Prior to coming to USDA, Johanns was Nebraska's 38th governor. During his six years in office, Johanns was a strong advocate for rural communities and farmers and ranchers. That's why, with a new farm bill on the horizon, Johanns went to the country in 2005 to hear firsthand from producers about what was working with current farm policy and what was not. Johanns hosted 21 of 52 farm bill forums held in 48 states.

Secretary Johanns is a graduate of St. Mary's University of Minnesota in Winona. He earned a law degree from Creighton University in Omaha and practiced law in O'Neill and Lincoln, Nebraska. Johanns served on the Lancaster County Board from 1983 to 1987, and on the Lincoln City Council in 1989–1991. He was elected mayor of Lincoln in 1991. He was reelected in 1995, and successfully ran for governor three years later. Secretary Johanns is married to Stephanie Johanns, a former Lancaster County Commissioner and State Senator. The couple has two children and three grandchildren.

Nicholas R. Jordan is a Professor in the Department of Agronomy and Plant Genetics at the University of Minnesota, and Director of Graduate Studies for the Sustainable Agricultural Systems Graduate Minor Program. His research interests include ecology of plant invasion, participatory development of integrated weed management methods, and ecology, management, and development of diversified "multifunctional" agricultural landscapes that produce ecological services and agricultural commodities. He is also interested in combining scientific knowledge with other ways of knowing to create an adequate knowledge base for sustainable agriculture. Currently, he is working to help organize and conduct participatory action research with coalitions of social groups in support of market development for the production of diversified and multifunctional agriculture. His teaching responsibilities include courses on agricultural ecology and systems thinking in sustainable agriculture. Dr. Jordan received his Ph.D. in botany and genetics from Duke University and his B.S. in biology from Harvard College.

Wynetta Y. Lee is the associate vice president for academic planning, research, and graduate studies at California State University–Monterey Bay. She has a successful career as a faculty member and as a leader in higher education. She served as an associate professor of higher education in the Department of Adult and Community College Education at North Carolina

State University. She is best known for her research on micropopulations, policy/program impact, and student performance. Her research addresses issues such as educational parity, mentoring, college student transfer, student development, and the disparity effect of policy/practice on institutions of color. Her publications and assessment reports reflect her broad interest in student achievement, educational equity, outcomes-based educational assessment, institutional policy, and the assessment of academic quality. She is a frequent contributor to knowledge through various chapters, monographs, and articles in the higher education literature. She is a member of the Association for the Study of Higher Education and editor of its *Assessment & Evaluation* literature, the American Educational Research Association, the Postsecondary Preparation Working Group, and the National Postsecondary Education Cooperative. Although she has a well-developed reputation as a researcher, teacher, and leader in higher education, Lee is most proud of those for whom she has been honored to serve as mentor into higher education careers.

Michael V. Martin. See committee biographies in Appendix F.

Jose P. Mestre is a professor of physics and educational psychology at the University of Illinois at Urbana-Champaign. His research interests include cognitive studies of problem solving in physics with a focus on the acquisition and use of knowledge by experts and novices. Most recently, his work has involved investigating transfer of learning in science problem-solving, applying research findings to the design of instructional strategies that promote active learning in large physics classes, and developing physics curricula that promote conceptual development through problem-solving. He has served on the National Research Council's Mathematical Sciences Education Board, and Committee on Developments in the Science of Learning; the College Board's Sciences Advisory Committee, SAT Committee, and Council on Academic Affairs; the Educational Testing Service's Visiting Committee, and Graduate Research Examination Technical Advisory Committee; the American Association of Physics Teacher's Research in Physics Education Committee and the editorial board of *The Physics Teacher*; and the Expert Panel of the Federal Coordinating Council for Science, Engineering and Technology. He has published numerous research and review articles on science learning and teaching, and has co-authored or co-edited 17 books.

Jay Moskowitz is Associate Vice President for Health Sciences Research, Vice Dean for Research & Graduate Studies of the College of Medicine at

Pennsylvania State University. He is also the Chief Scientific Officer of the Milton S. Hershey Medical Center. Dr. Moskowitz has 27 years of experience at the National Institutes of Health where he started his career as a Postdoctoral Fellow in the Pharmacology Research Associate Program, National Institute of General Medical Sciences, and went on to serve as Principal Deputy Director of the National Institutes of Health (NIH). Dr. Moskowitz dedicated his NIH career to developing programs that would serve to facilitate the research careers of emerging basic and physician investigators. He was responsible for developing the Pulmonary Young Investigator Award, Pulmonary Academic Award, numerous trans-NIH career development K awards, and the Shannon Award. He spent eight years between his appointment at NIH and Penn State at the Wake Forest University School of Medicine as Senior Associate Dean for Science and Technology. Dr. Moskowitz received his Ph.D. from Brown University, and his B.A. from Queens College, City University of New York.

Jeanne L. Narum is the founding director of Project Kaleidoscope, an informal national alliance that focuses on building leadership at the institutional and national levels to ensure that American undergraduates have access to robust learning experiences in science, technology, engineering, and mathematics. Jeanne is also the director of the Independent Colleges Office, which assists its member institutions in being competitive in their search for grants from federal agencies for faculty and curriculum development and institutional renewal. She previously served as director of government and foundation relations at St. Olaf College, director of development at Dickinson College, and vice president for development and college relations at Augsburg College. She has served on several National Research Council committees on undergraduate education in the sciences. She has received honorary doctorates from the University of Portland, Ripon College, and the University of Redlands. Jeanne received her Bachelor of Music from St. Olaf College.

Marion Nestle is the Paulette Goddard Professor in the Department of Nutrition, Food Studies, and Public Health at New York University, which she chaired from 1988 to 2003. She has held faculty positions in the Department of Biology at Brandeis University and at the University of California, San Francisco, School of Medicine, where she was Associate Dean for Human Biology Programs. She was the senior nutrition policy advisor in the Department of Health and Human Services and managing editor of the 1988 *Surgeon General's Report on Nutrition and Health*. She was a member of the FDA Food Advisory Committee and Science Board, the USDA/DHHS 1995 Dietary Guidelines Advisory Committee, and American Cancer Society committees that issue dietary guidelines. She is currently a member of the Pew Commission on Industrial Farm Animal Production. She is the author of *Food Politics: How the Food Industry Influences Nutrition and Health* (2002), which won awards from the Association for American Publishers, James Beard Foundation, and World Hunger Year, and author of *Safe Food: Bacteria, Biotechnology, and Bioterrorism* (2003), which won NYU's Griffiths Research Award and was selected as a 2004 Best Book by the *San Francisco Chronicle.* In 2004, she was named alumna of the year by the University of California School of Public Health, and received the David P. Rall Award for Advocacy in Public Health from the American Public Health Association. In 2005, she was elected as a Fellow of the American Society for Nutritional Sciences and of the American Association for the Advancement of Science; and received the Health Quality Award from the National Committee for Quality Assurance and the Bridging the Gap Award for Excellence in Science and Public Policy Writing from the Northern California Public Health Association. Her latest book, *What to Eat,* was published in May 2006. Dr. Nestle completed a Ph.D. in molecular biology and an M.P.H. in public health nutrition from the University of California, Berkeley.

James L. Oblinger. See committee biographies in Appendix F.

H. Paul Roberts is the director of study abroad and international programs at the College of Agriculture and Natural Resources at Michigan State University (MSU). He has been involved in international programs at MSU for almost 30 years, having served as Assistant to the Vice Provost and Dean, Acting Associate Dean for International Program, and Director of Study Abroad and International Training for the College of Agriculture and Natural Resources. Under his guidance, the College has developed the largest study abroad program in agriculture in the United States with more than 50 programs in 30 countries. He has personally conducted more than 30 international programs involving more than 500 students. He received the 2005 MSU award for outstanding service to study abroad. Dr. Roberts also teaches courses on "Global Issues in Agriculture and the Environment" on campus.

Gary Rodkin is Chief Executive Officer of ConAgra Foods, Inc. Prior to joining the company in 2005, Mr. Rodkin was Chairman and CEO of PepsiCo Beverages and Foods North America, where he led a $10 billion organization including such leading brands as Pepsi, Gatorade, Quaker Foods, and Tropicana. He joined PepsiCo in 1998 when PepsiCo acquired Tropicana, where he had served as its president since 1995. From 1979 to 1995, Mr. Rodkin held marketing and general management positions of increasing responsibility at General Mills, participating in the successes of many of its leading brands from Cheerios to Betty Crocker, with his last three years at the company as president, Yoplait-Colombo. Mr. Rodkin earned a bachelor's degree in economics from Rutgers College and a MBA from Harvard Business School.

Sally L. Shaver is the Associate Counselor for Agricultural Policy in the Office of Air and Radiation at the U.S. Environmental Protection Agency. Sally Shaver has over 33 years of government experience. She began her career at the South Carolina Department of Health and Environmental Control where she developed and worked on water quality models. Her career with the EPA began in the regional office in Atlanta, Georgia, where she worked in all aspects of the water program and was the lead for permitting in the air program before moving to the Agency for Toxics Substances and Disease Registry at the Centers for Disease Control and Prevention where she worked with the Superfund program. From there she moved back to the EPA in Research Triangle Park, North Carolina, where she was responsible for setting and implementing the national ambient air quality standards for several years before spending eight years in charge of the air toxics program. She has led the U.S. delegation on two international task forces and has been a member of USDA's Agricultural Air Quality Task Force since its inception. Ms. Shaver has a B.S. in mathematics from Furman University and an M.S. in environmental engineering from Clemson University.

Paul D. Tate is a Senior Scholar in Residence at the Council of Graduate Schools. He is a co-director of the Professional Science Master's Initiative, funded by the Ford Foundation and by the Alfred P. Sloan Foundation. He is also a professor of philosophy at Idaho State University and a former Dean of Graduate Studies at Idaho State University, where he continues to work on a special project to develop a training program in research ethics for graduate students. Dr. Tate studied in India and Sri Lanka on a Fulbright scholarship and taught in Sri Lanka as a Fulbright scholar. In addition to scholarly articles on early Sanskrit literature and on the German philosopher Martin Heidegger, Dr. Tate has published several works of fiction, all set in South Asia. In 1996 he organized a conference in India on ethical and political issues in cross-cultural art—issues he continues to address in his literary work and in his collaborations with visual artists. Dr. Tate is a founding member of the ethics committee of the Portneuf Medical Center, and he offers frequent workshops on ethics in medicine, business, engineering, research, and university administration. Dr. Tate received his Ph.D. and M.Phil. in philosophy from Yale University, and his B.A. in philosophy from University of Texas at Austin.

Andrew L. Waterhouse is the John E. Kinsella Chair in Food, Nutrition, and Health and a professor of enology at the University of California, Davis. His research focuses on the chemistry of phenolic compounds and addresses two types of effects: the taste of wine and health effects of wine to consumers. In both cases, his lab collaborates with others who can help utilize chemical data and assistance to advantage and vice versa. In the area of wine quality, his interest is in the effect of oxidation on wine chemistry and how this oxidation affects important quality parameters of wine, such as taste and color. Dr. Waterhouse has been studying micro-oxygenation and its effect on wine color and tannins, and is currently testing some new theories on wine oxidation chemistry. He also participates in the development of general analytical methodology of interest in wine analysis, has published a few different methods in this area, and is applying a number of different methods to look at new grape or wine treatments being offered by various companies. Dr. Waterhouse received his Ph.D. in chemistry from the University of California, Berkeley, and his B.S. in chemistry from the University of Notre Dame.

Robin Wright is Associate Dean for Faculty and Academic Affairs in the College of Biological Sciences (CBS) and professor of Genetics, Cell Biology, and Development at the University of Minnesota. Her lab studies the genetic control of cell structure, using yeast as a model organism. In her previous position at the University of Washington, she taught nonmajors' biology and introductory and advanced cell biology. Wright spends considerable effort on activities that promote innovation and improvement of undergraduate education. Her teaching effectiveness was recognized by a University of Washington Distinguished Teaching Award in 2000. At the University of Minnesota, she chairs the CBS Curriculum Task Force as well as the university's Council on Enhancing Student Learning. In addition to teaching freshman seminars, an honors colloquium, and introductory biology, she also helped to develop and co-teaches an orientation/enrichment course required for all incoming freshmen in the college.

Robert T. Yuan is a part-time senior staff officer at the Board on Life Sciences at the National Research Council and is a professor emeritus in Cell Biology and Molecular Genetics at the University of Maryland, College Park. During his 19 years at the University of Maryland, he was a co-Principal Investigator of the East Asia Science and Technology Project which introduced East Asian themes into undergraduate science and engineering courses. He created three honors seminars and one senior-level microbial physiology course and worked with faculty teams to create an honors seminar and completely restructure the required general microbiology course. Dr. Yuan was a founder of a biotechnology company that focused on drug discovery, and he established a biotechnology consulting group that provided services to foreign governments, private companies, and financial organizations. He was also a U.S. Foreign Service officer that carried out an assessment of biotechnology in Western Europe while he was based at the U.S. Embassy in London. Previous to that he was a section chief at the National Cancer Institute and had done research and taught at Harvard University, Edinburgh University (UK), and Basel University (Switzerland). Dr. Yuan completed his Ph.D. at the Albert Einstein College of Medicine.

PARTICIPATING INSTITUTIONS AND ORGANIZATIONS

The list below includes the institutional affiliation of those who were registered to participate in the Leadership Summit. Actual participation may vary slightly.

- Association of American Veterinary Medical Colleges
- AgCareers.com
- AgrowKnowledge
- Alabama A&M University
- Arkansas State University
- American Agricultural Economics Association
- American Society of Agronomy
- Auburn University
- Biotechnology Institute, The
- California Polytechnic State University, San Luis Obispo
- California State University, Monterey Bay
- Cargill, Inc.
- Carleton College
- Colorado State University
- ConAgra Foods, Inc.
- Cornell University
- Council on Food, Agricultural and Resource Economics, The
- Council of Graduate Schools
- Crop Science Society of America
- Dow AgroSciences LLC
- Farm Foundation
- Florida A&M University
- Food Systems Leadership Institute
- General Mills, Inc.
- Georgia Institute of Technology
- Grambling State University
- Haskell Indian Nations University
- Hispanic Association of Colleges and Universities
- Howard Hughes Medical Institute
- Illinois State University
- Institute of Food Technologists
- International Food and Agribusiness Management Association
- Interuniversity Consortium for Agricultural and Related Sciences in Europe

- Iowa State University
- Jasper Wyman & Son
- Kansas State University
- Kirkwood Community College
- Michigan State University
- Mississippi State University
- National Academies, The
- National Association of State Universities and Land-Grant Colleges
- National Science Foundation
- New Mexico State University
- New York University
- North American Colleges and Teachers of Agriculture
- North Carolina A&T State University
- North Carolina State University
- North Central Regional Association of Agricultural Experiment Station Directors
- Northeastern Regional Association of Agricultural Experiment Station Directors
- Ohio State University, The
- Oklahoma State University
- Oregon State University
- Pennsylvania State University
- Pioneer Hi-Bred International, Inc.
- Project Kaleidoscope
- Purdue University
- Rutgers University, The State University of New Jersey
- Soil Science Society of America
- South Dakota State University
- Southern Association of Agricultural Experiment Station Directors
- Strategic Education Research Partnership Institute
- Texas A&M University
- Texas Tech University
- University College Dublin, Ireland
- University of Arkansas
- University of California System
- University of California, Davis
- University of Connecticut
- University of Florida
- University of Georgia
- University of Idaho

- University of Illinois at Urbana-Champaign
- University of Kentucky
- University of Maryland, College Park
- University of Minnesota, Twin Cities
- University of Missouri–Columbia
- University of Natural Resources and Applied Life Sciences, Vienna, Austria
- University of Nebraska–Lincoln
- University of New Hampshire
- University of Puerto Rico
- University of Rhode Island
- University of Tennessee–Knoxville
- University of the District of Columbia
- University of Wisconsin–Madison
- University of Wyoming
- U.S. Department of Agriculture
- U.S. Department of Education
- U.S. Department of Energy
- U.S. Environmental Protection Agency
- U.S. House of Representations
- Utah State University
- Virginia Polytechnic Institute and State University
- Walt Disney World, Epcot Center
- West Texas A&M University
- West Virginia University
- Western Association of Agricultural Experiment Station Directors
- Wilmington College

Appendix C

Shifts in the Production and Employment of Baccalaureate Degree Graduates from United States Colleges of Agriculture and Natural Resources, 1990-2005[1,2]

Background Paper by:
Jeffrey L. Gilmore (U.S. Department of Agriculture [USDA]),
Allan D. Goecker (Purdue University),
Ella Smith (USDA),
P. Gregory Smith (USDA)

Contributors:
Franklin E. Boteler (USDA),
Jorge A. González (U. Puerto Rico-Mayagüez),
Joe Hunnings (University of Vermont),
Timothy P. Mack (Virginia Polytechnic Institute and State University),
A. Dale Whittaker (Purdue University)

INTRODUCTION

This paper will highlight some of the major trends characterizing the milieu in which agricultural higher education has operated over the past 15 years, including an examination of the shifts in student demographics, graduation and degree patterns, employment opportunities, college structure and majors, the business and social environment, and consumer preferences. In order to better examine the current state of affairs, it might be helpful to

[1]This report draws heavily on material from national data collected by the U.S. Department of Education, the U.S. Department of Labor, the National Science Foundation, the U.S. Census Bureau, and the U.S. Department of Agriculture. The analyses and views expressed here, and any attendant errors or omissions, are the sole responsibility of the authors and do not represent the positions or policies of their employing agencies or the National Academy of Sciences.

[2]This paper has been updated from its original version to incorporate data made available since the 2006 Leadership Summit. The only changes are to add more recent data to several figures and update the text references to those data.

first provide a quick review of agricultural higher education history and the involvement of the U.S. Department of Agriculture (USDA).

The 1st Morrill Act of 1862 established the land-grant system to provide for a "practical education" in agriculture and the mechanical arts for the common man. This was in stark contrast to the existing system of private colleges for the elites, which provided training for lawyers, physicians, and the clergy. In those days most people lived on farms, and the "Ag School" was the core of the new land-grant colleges. Not coincidentally, 1862 also saw the establishment of the USDA as "The Peoples' Department" to serve rural America.

Shifting forward 115 years to 1977, the situation had changed dramatically. The land-grant colleges, including the 1862 and 1890 institutions, had evolved into world-class universities, but colleges of agriculture were no longer the entire university, or even a core unit in many cases. Other public and private institutions, including community colleges, became involved in the education of students in the fields of agriculture though not at the same breadth and depth as the land-grant institutions. Leaders of America's agricultural higher education programs requested Congress to transfer the lead federal role for facilitating agricultural higher education programs from the U.S. Office of Education to USDA. It was felt that agricultural and natural resources higher education programs could be conducted more effectively in concert with the USDA's agricultural research and extension programs. As a result of these efforts, agricultural higher education program authority was transferred to the USDA in provisions of the 1977 Farm Bill.

Since then, there have been a number of developments. In implementing congressional authorities and appropriations, USDA established a National Needs Graduate Fellowships program for scientific human capital development in 1984, and in 1990 USDA initiated the Higher Education Challenge Grants program to modernize food, agricultural, and natural resources curricula, improve instructional delivery systems, stimulate student recruitment and retention, encourage faculty development, and expand student experiential learning opportunities.

In 1988, the USDA sponsored a national summit focusing on graduate education in agriculture. In April 1991, the National Research Council's Board on Agriculture held the first conference on higher education to "chart the comprehensive changes needed to meet the challenges of undergraduate professional education in agriculture." Topics of papers and discussions included the core curriculum, diversity and multiculturalism, scientific literacy, undergraduate research, rewarding teaching excellence, globalism, curricular innovation, agriculture as a science, and the science of agriculture.

USDA now invests over $100 million annually in higher education programs through 20 national initiatives that help support agricultural and natural resources colleges both within and outside the land-grant college system. During the past quarter-century, agricultural and natural resources curricula have been transformed to challenge and serve students with broadening professional interests and academic backgrounds. Facilities and equipment have been modernized to incorporate contemporary information technologies and biotechnologies. Increased emphasis is now being placed on active learning methodologies and experiential education, including undergraduate research, internships in the public and private sectors, and study abroad opportunities. Outstanding students have been attracted to graduate study in agricultural and natural resources via graduate fellowships, and faculty recognition programs for outstanding teaching have been initiated. In addition, many colleges have changed their identities from a focus limited to agriculture to one emphasizing a broader scope of study, while other colleges have entirely eliminated a reference to agriculture in their names. As the lead federal agency for agricultural and natural resources higher education programs, USDA has worked successfully with the nation's colleges and universities to transform programs of study and generate graduates with new and contemporary skills and attributes.

It is against this backdrop that we examine the evolving characteristics of graduates having expertise in food, agricultural, and natural resources disciplines, and set this examination within the current context of changing professional opportunities to meet the human resources needs of employers. It is an exciting and rapidly shifting paradigm requiring careful analyses, visionary thinking, and decisive actions.

TRENDS IN BACCALAUREATE DEGREES AWARDED BY COLLEGES OF AGRICULTURE AND NATURAL RESOURCES

Significant growth in the number of agricultural and natural resources baccalaureate degree recipients occurred in the United States between 1987 and 2007. In the 1987–88 Academic Year (AY), colleges and universities awarded 18,572 baccalaureate degrees in agricultural and natural resources disciplines compared to 33,680 in AY 2006–07. Much of the growth in degrees conferred, as reported by the National Center for Educational Statistics, was realized in three areas of study, including Natural Resources Conservation and Research, Animal Sciences, and Agricultural Business and Management.

Figure C-1 shows that 872 baccalaureate degrees were awarded in Natural Resources Conservation and Research in AY 1987–88 compared

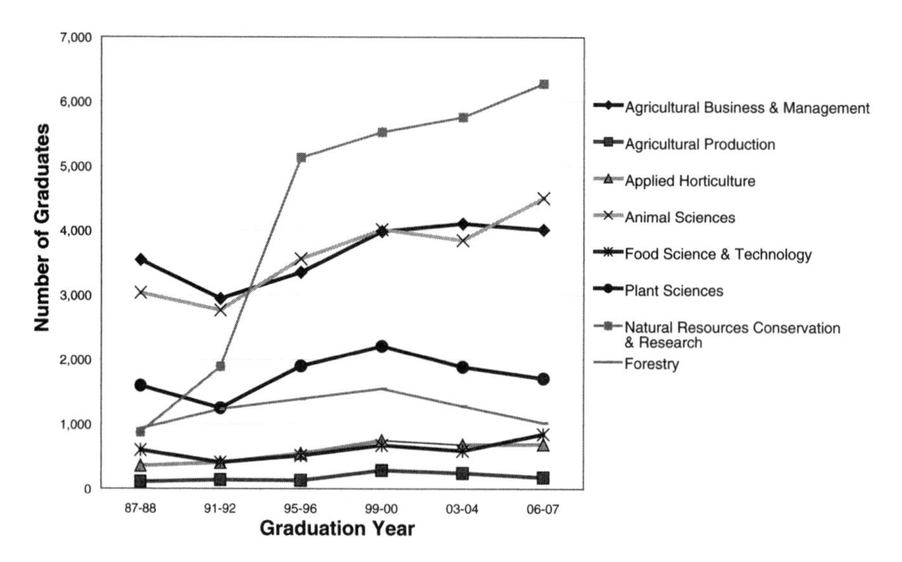

FIGURE C-1 Number of baccalaureate degrees awarded in selected agricultural and natural resources fields of study, United States, 1989–2007.
SOURCE: National Center for Education Statistics Completion Reports. [This figure has been updated to include data made available since the Leadership Summit.]

to 6,276 in AY 2006–07. Animal Sciences baccalaureate degrees increased from 3,034 to 4,505 during this time, while Agricultural Business and Management degrees rose from 3,542 to 4,010.

During the period between 1987 and 2007, baccalaureate degrees in Agricultural Production increased from 109 in AY 1987–88 to 177 in AY 2006–07. Applied Horticulture degrees rose from 356 to 688. Plant Sciences degrees increased from 1,592 to 1,706 and Forestry degrees went up from 930 to 1,019.

While there was significant expansion in the aggregate number of degrees awarded between 1987 and 2000, the number of degrees awarded after 2000 begins to stabilize.

AGRICULTURAL AND NATURAL RESOURCES BACCALAUREATE DEGREES AS COMPARED TO ALL BACCALAUREATE DEGREES

During the period between 1987 and 2007, the number of baccalaureate degrees awarded in agricultural and natural resources areas of study increased by 80 percent. In contrast to this, baccalaureate degrees awarded in all areas of study increased by only 60 percent (Figure C-2).

Most of the growth in the number of agricultural and natural resources baccalaureate degrees occurred in the mid and late 1990s, and reflected steep enrollment increases experienced by the nation's colleges of agriculture and natural resources in the late 1980s and early 1990s. (As previously noted, enrollments in these areas either remained stable or declined in recent years.) In comparison, baccalaureate degrees awarded in all fields of study in the United States continued to increase throughout the period from 1987 to 2007.

DEMOGRAPHIC CHARACTERISTICS OF GRADUATES IN AGRICULTURAL AND NATURAL RESOURCES FIELDS OF STUDY

During the period from 1987 to 2007, the number of baccalaureate degrees in the agricultural and natural resources fields of study awarded to females rose significantly from 6,284 in AY 1987–88 to 16,262 in AY 2006–07. During the same period, baccalaureate degrees awarded to males increased from 12,288 in AY 1987–88 to 17,509 in AY 1999–2000, but declined to 17,418 by AY 2006–07. These data are depicted in Figure C-3.

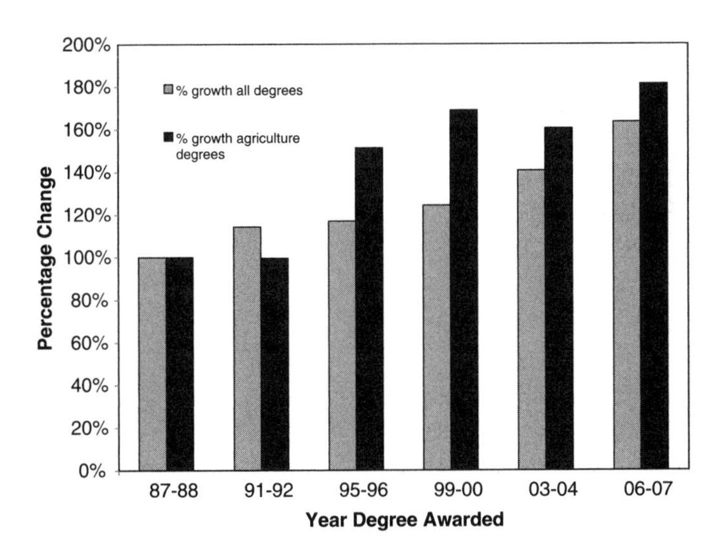

FIGURE C-2 Index of relative growth in bachelor degrees awarded in selected agricultural specialties compared to all bachelor degrees awarded at U.S. institutions, 1987–2007.
SOURCE: National Center for Education Statistics Completion Reports. [This figure has been updated to include data made available since the Leadership Summit.]

During the period from 1995 to 2007, there was relatively little change in the racial and ethnic characteristics of baccalaureate degree recipients in agricultural and natural resources programs of study. In AY 1995–96, a little over 87 percent of the graduates were White non-Hispanic compared to 81 percent in AY 2006–07. Black non-Hispanic graduates increased very little, from 2.9 to 3.2 percent, American Indian/Alaska Native from 0.7 to 0.8 percent, Asian or Pacific Islander from 2.4 to 4.4 percent, and Hispanic from 2.7 to 4.6 percent.

As Figure C-4 shows, in AY 2003–04 a total of 1,089 agricultural and natural resources baccalaureate degrees were awarded to Black non-Hispanics, 272 to American Indian/Alaska Native, 1,464 to Asian or Pacific Islander, and 1,547 to Hispanic populations. The remainder of the 27,281 degrees awarded went to White non-Hispanic students. While overall numbers have not changed much, a significant development is in the number of Hispanic graduates, which has increased and recently surpassed the number of Black non-Hispanics and Asian or Pacific Islanders. The number of nonresident aliens and American Indian/Alaska Natives has remained constant.

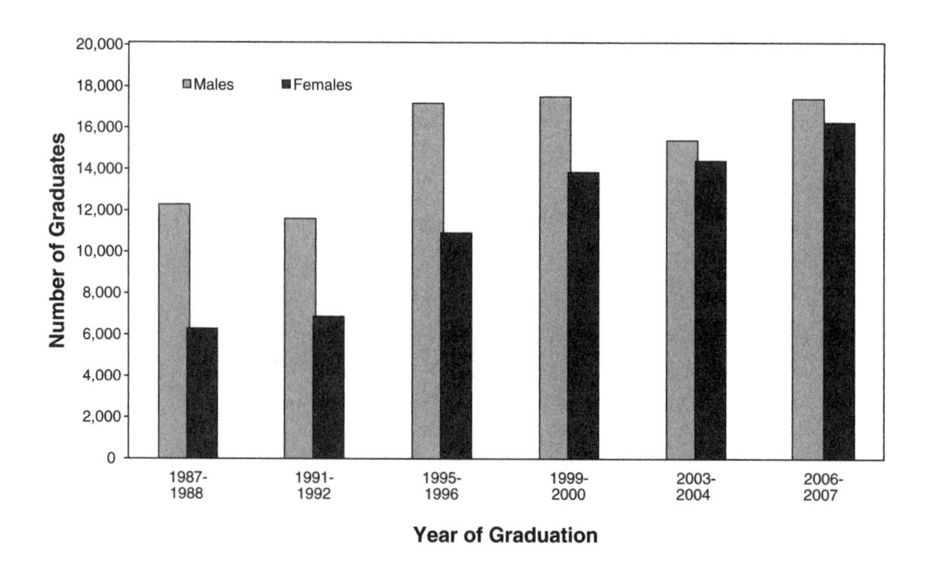

FIGURE C-3 Gender of baccalaureate degree recipients in selected agricultural and natural resources degree fields of study, United States, 1989–2007.
SOURCE: National Center for Education Statistics Completion Reports. [This figure has been updated to include data made available since the Leadership Summit.]

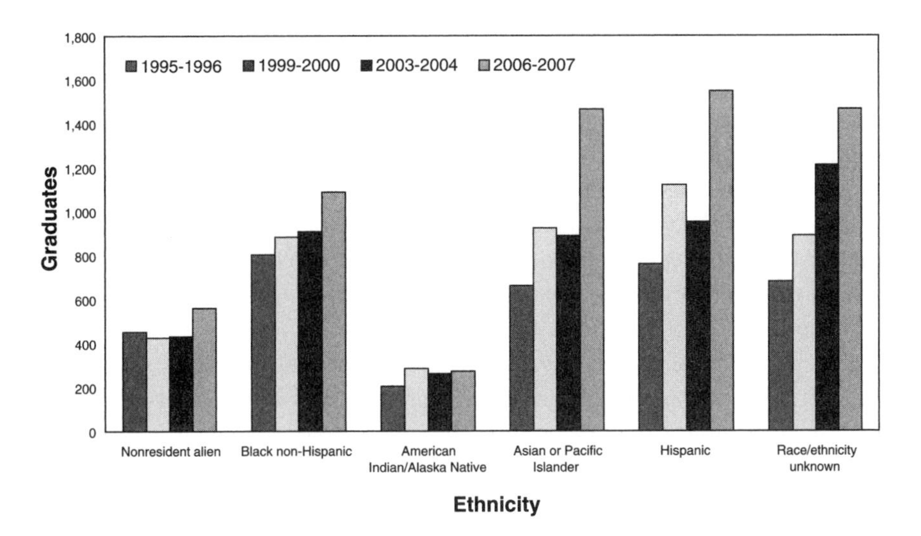

FIGURE C-4 Number of baccalaureate degrees awarded in agricultural and natural resources fields of study by selected ethic groupings, United States, 1995–2007.
SOURCE: National Center for Education Statistics Completion Reports. [This figure has been updated to include data made available since the Leadership Summit.]

As Table C-1 shows, somewhat greater variations in demographic characteristics are observed between the degree levels in agriculture, natural resources, and veterinary medicine specializations.

TABLE C-1 Selected Demographic Characteristics of Graduates in Agriculture, Natural Resources, and Veterinary Medicine Fields of Study, United States, 2001–2002

Degree Level	Females (%)	Ethnic Minorities (%)	Non-U.S. Citizens (%)
Baccalaureate	53	16	2
Master's	55	14	15
Doctor of Philosophy	41	17	35
Doctor of Veterinary Medicine	72	9	1

SOURCE: National Center for Education Statistics Completion Report 2001–2002.

PROJECTED AVERAGE ANNUAL EMPLOYMENT OPPORTUNITIES AND AVAILABLE GRADUATES IN AGRICULTURAL AND NATURAL RESOURCES

During the past three decades, a series of five-year studies has been sponsored by the Higher Education Programs unit of the USDA Cooperative State Research, Education, and Extension Service. The purpose of these studies is to project and compare the number of qualified college graduates that are available to fill the expected number of employment opportunities requiring expertise in food, agricultural, and natural resources specialties.

Summary data from the four most recent studies are presented in Figure C-5. These graphs are based upon analyses of Bureau of Labor Statistics and Department of Education data, and show projected job openings in agricultural and natural resources occupations (broadly defined) compared to projected numbers of qualified graduates from 1990 to 2010. Strong U.S. economic conditions in the late 1990s, when the 2000–2005 projections were developed, contributed to the relatively higher number of projected employment opportunities during the period.

Two sources of graduates with requisite expertise in agricultural and natural resources specialties have been utilized to project the average annual availability of qualified graduates charted in Figure C-5. "Agriculture degree recipients" are the baccalaureate, master's, doctoral, and doctor of veterinary medicine degree graduates generated by colleges of agriculture and natural resources, and by colleges of veterinary medicine. "Allied degree

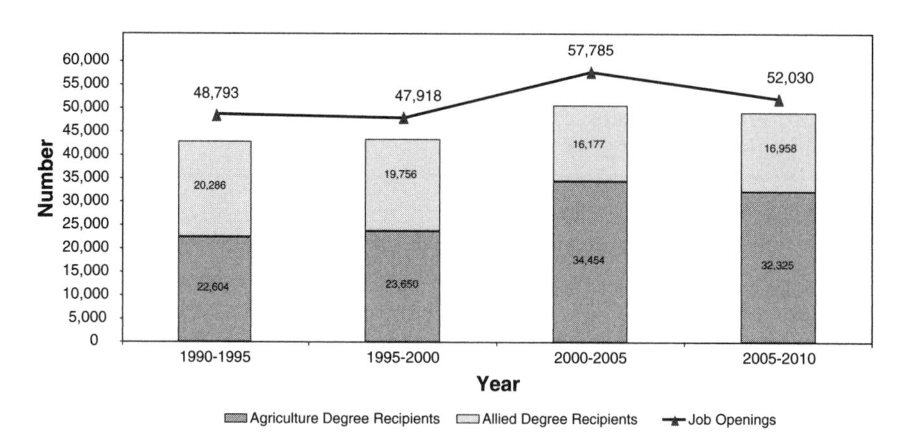

FIGURE C-5 Projected average annual employment opportunities and available graduates in agricultural and natural resources fields of study, United States, 1990–2010. SOURCE: U.S. Department of Labor Monthly Labor Review, February 2004, and National Center for Education Statistics Completion Reports.

recipients" are graduates of other academic units, including colleges of engineering, arts and sciences, and business schools, who are deemed to have the requisite expertise necessary to fill job openings in agricultural and natural resources occupations. It is important to note (as stated above) that graduates at all degree levels, not just baccalaureate degrees recipients, are included in the Figure C-5 analyses.

Projected areas of employment strengths and weaknesses for 2005 to 2010 are discussed below.

Management and business occupations: Strong employment opportunities are expected for technical sales representatives, accountants and financial managers, market analysts, landscape managers, and international business specialists. Weaker employment opportunities are forecasted for sales and business representatives who provide services to farmers and ranchers, and grain and food animal merchandisers.

Scientific and engineering occupations: Most employment opportunities are expected for graduates with skills in precision agriculture, functional genomics and bioinformatics, forest science, plant and animal breeding, biomaterials engineering, food quality assurance, nanotechnology, animal health and well-being, nutraceuticals development, and environmental science. Expect relatively fewer opportunities for agricultural machinery engineers, wildlife and range scientists, and veterinarians in general practice.

Agricultural and forestry production occupations: Good job opportunities are projected for producers of fruits and vegetables, growers of specialty crops that provide raw materials for medical and energy products, managers of specialized livestock operations, forest resources managers, growers of landscape plants and trees, managers of aquaculture operations, turf producers, organic farmers, and providers of outdoor recreation. However, as agricultural production units continue to consolidate, there will be fewer opportunities for producers of traditional commodities (e.g., wheat, corn, cotton, soybeans, cattle, and hogs).

Education, communication, and governmental services occupations: Most opportunities are projected in plant and animal inspection, public health administration, biotechnology impact assessment, nutritional and health occupations geared to serve an aging population, outdoor recreation, food system security, consumer information technologies, and environmental and land-use planning. More limited opportunities will be found for farm and ranch advisors, and government farm service agents.

Results of the most recent study are available at <http://faeis.usda.gov/supplydemand/2005-2010/>.

Some significant assumptions regarding socioeconomic forces and anticipated technological advancements must be factored into the model to project employment opportunities for graduates. What follows is a discussion of the four factors that were considered to be most important when generating the projections for 2005–2010. The factors are:

(1) Consumers and their preferences dictate that products and services derived from agricultural and forest raw materials must help them maintain contemporary lifestyles. Population growth, changing ethnic and age demographics, and evolving food and health literacy strongly influence both what is produced and the expertise required to meet consumer demands.

(2) The evolving business structures that support the U.S. food system continue to be influenced by globalization and consolidation. Expertise needs will evolve and create a need for graduates with excellent business skills, international understanding, and leadership qualities. Graduates must deal with increasing market uncertainty, risk analysis, petroleum dependence, niche business opportunities, and global food production and distribution systems.

(3) New developments in science and technology are being driven by changes in biosecurity, the expanding global population, health concerns, shrinking natural resources, and climate change. Emerging biotechnologies and nanotechnologies are powerful tools to increase food system efficiency. Other scientific developments will help us maintain our renewable natural resources. All of these require graduates with basic science skills and the ability to solve problems with scientific applications.

(4) Public policy choices and accountability will affect the market for graduates who can provide public services, including education, natural resource utilization, food assistance, recreation, and financial support. Public concerns regarding diet and health, food safety, and the environment dictate the number and kinds of graduates needed to manage regulatory programs and provide services to assist producers and others working in the food and natural resource system.

IMPORTANT FACTORS AFFECTING FUTURE AGRICULTURAL AND NATURAL RESOURCES GRADUATES

Many factors are impacting higher education institutions as they offer academic programs to prepare future graduates in the agricultural and natural resources sciences. The factors that are especially important include the racial and ethnic characteristics of K–12 students, student and family misconceptions about agriculture careers, and the changing skill sets employers seek. These factors are discussed below.

Figure C-6 indicates the demographic trends in the racial and ethnic composition of students in U.S. public schools. A steady increase in the percentage of minority students over the last 30 years is shown (22 percent in 1972 compared to 39 percent in 2002) with the percentage of Hispanic students increasing from 6 percent to 18 percent over the same time period. While agricultural and natural resources higher education programs have been working to attract more minority students, there have only been very small increases in minority baccalaureate degree recipients from 1995 to 2004. Diversity continues to be a major opportunity and challenge to colleges of agriculture and natural resources.

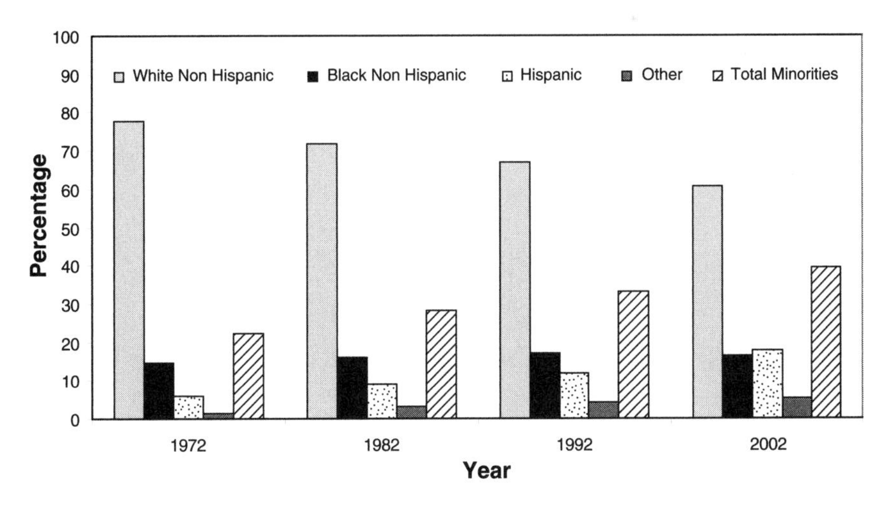

FIGURE C-6 Racial/ethnic distribution of public schools, grades K–12, United States, 1972–2002.
SOURCE: US Department of Commerce, Census Bureau, Current Population Survey, 1972–2004.

TABLE C-2 Main Concerns Affecting U.S. High School Students in Selecting Agricultural Sciences as a Career Major

Main Concern	Percentage
Misconception or Image about Agricultural Sciences	41
Lack of Knowledge about Employment Opportunities	33
Lack of Knowledge about Fields of Study	22
Perceived Relevance/Importance to Future Career	22
Students Lack Fundamental Knowledge in Mathematics and Sciences	11
Peer Pressure/Family Against Agricultural Sciences Studies	7

SOURCE: Gonzalez 2006.

In 2005, academic program administrators in colleges of agriculture and natural resources evaluated the factors affecting student choice to seek admission and matriculate. Results of the survey are presented in Table C-2.

These data suggest that colleges of agriculture and natural resources continue to be challenged in helping potential students better understand the academic and career opportunities in these fields. In addition, there appears to be continuing reason for concern regarding the public's perception of the images associated with agricultural and natural resources programs of study.

Table C-3 presents the skills that agribusiness employers have identified as the ones most important for new college graduates. Colleges of agriculture and natural resources must continually update courses and curricula to meet changing expectations in the employment arena. Portfolios of faculty and academic resources may or may not be positioned to offer academic programs capable of generating graduates with the high-priority skills and preparation that employers seek.

EVOLVING HIGHER EDUCATION PROGRAMS IN AGRICULTURAL AND NATURAL RESOURCES

Agricultural and natural resources colleges have responded to these concerns by consolidating and realigning their offerings with other programs, and by changing their names and structure. For example, at the 58 traditional 1862 land-grant institutions, 49 have an agricultural college. Of these, 12 are named the "College of Agriculture" while 37 have names encompassing agriculture along with something else, most commonly natural resources, life sciences, environmental sciences, food sciences, biological sciences, or family and consumer sciences. These changes are reflected at non-land-grant

TABLE C-3 Skill Sets and Abilities that Agribusiness Employees Seek in New College Graduates

Skill Sets and Abilities	Rating[a]
Interpersonal Communication Skills	5.00
Critical Thinking Skills	4.92
Writing Skills	4.36
Computer Skills	4.27
Cultural/Gender Awareness/Sensitivity	4.08
Quantitative Analysis Skills	4.07
Knowledge of Business Management	4.00
Oral Presentation Skills	4.00
Knowledge of Accounting and Finance	3.62
Intern/Co-op Work Experience	3.29
Knowledge of Macroeconomics, International Trade	3.08
Broad-based Knowledge in Liberal Arts	2.75
International Experience	2.75
Foreign Language Skills	2.56
Production Ag Experience	2.36

[a]Rated on scale of 1 to 5, where 1 is unimportant and 5 is absolutely essential.
SOURCE: Adapted from Boland and Akridge 2006.

institutions as well. Names are trending toward the life sciences, and as a consequence, the public image of "agriculture" is broadening.

Along with the name changes for colleges of agriculture, departments within the colleges also are shifting. The traditional food, agricultural sciences, and natural resources disciplines now also include biology, rangeland, statistics, communications, fisheries, parks and recreation, human development, and landscape architecture. Associate deans for academic programs at the 1862 land-grant institutions have recently projected the following majors as having the most growth potential: pre-veterinary science, equine/companion animal science, agricultural biotechnology, food science/food safety/nutrition, turf/landscape/urban horticulture, natural resources/environmental science, agribusiness, and families/communities/consumer sciences.

In contrast to the above fields of study, other traditional majors are projected to decline, including soil and crop science, entomology, animal science (meat animal), and plant pathology.

CLOSING THOUGHTS

What can we expect for the future? One recent study solicited responses to this very question. Results from that study suggest that agriculture's future will be filled with a host of new and emerging disciplines, including genomics;

genetics; molecular biology; computational biology; biological engineering/manufacturing; biosecurity; wellness; food/health interaction; human/animal interaction; animal behavior/wellbeing; renewable energy/resources; bio-based products; biosensors; biorenewable engineering; climate change; spatial sciences; water conservation, management, and policy; sustainable agriculture; land-use planning/policy; landscape restoration and design; human/environmental interaction; international/intercultural (agriculture/business); entrepreneurship; food production policy; health/science information and decision making; production/management/ecology of GMOs; and science/risk communication.

These emerging fields clearly reflect several societal changes—from producer to consumer, rural to suburban, and uninformed to educated. It appears, more and more, that "agriculture" is being defined as an area of basic sciences applied to wellness and sustainability. Is this our future? Will the land-grant institutions still be positioned to provide for a "practical education in agriculture and the mechanical arts for the common man?" Or, is this mission obsolete? We must not just "wait and see" but, rather, we must define and engineer the future we need and desire. Hopefully, the National Academy of Sciences Leadership Summit will do just that.

BIBLIOGRAPHY

Boteler, Franklin E. July 2006. Response to Recommendations of the National Food and Agribusiness Management Commission. Presentation at the American Agricultural Economics Association Meeting. Long Beach, California.

Goecker, Allan D., Jeffrey L. Gilmore, Ella Smith, and P. Gregory Smith. March 2005. *Employment Opportunities for College Graduates in the U.S. Food, Agricultural, and Natural Resources System, 2005-2010.* Purdue University College of Agriculture, West Lafayette, Indiana.

Gonzalez, Jorge A. August 2006. Agricultural Programs: Are They Able to Adapt for the Future? CSREES Faculty Fellow presentation at USDA. Washington, D.C.

Mack, Timothy P. November 2005. Lessons Learned about Agriculture by Doing FAEIS. Presentation at the NASULGC Annual Meeting, Academic Programs Section Workshop. Washington, D.C.

NASULGC. September 2005. *Directory: Deans and Directors of Academic Programs in Schools and Colleges of Agriculture, Agriculture and Life Sciences, or Agriculture and Natural Resources.* Washington, D.C.

Whittaker, A. Dale. November 2005. Trends in University Preparation for Agriculture and the Life Sciences. Presentation at the NASULGC Annual Meeting, Academic Programs Section Workshop. Washington, D.C.

Appendix D

Rethinking Undergraduate Science Education: Concepts and Practicalities—A Traditional Curriculum in a Changed World

Background Paper by:
Robert T. Yuan (University of Maryland, College Park, and National Research Council)

Science education at the university level has been based on a number of premises. Students that have successfully completed a course of study will have mastery of a scientific discipline. That knowledge should basically be sufficient to take them through a working life of about 40 years in a given career track, e.g., industrial research, and project area, e.g., mode of action of antibiotics. And that career takes place within the boundaries, physical and intellectual, of one country.

Let us then turn to the world we actually work and live in. Science and technology are interdisciplinary and most of the work is done by teams composed of individuals from different disciplines. The half-life of a project is likely to be on the order of seven years which means that an individual may have to retool him/herself several times in the course of a working life. It will also not be uncommon for that individual to have multiple career tracks, e.g., from academia to industry to venture capital. And new knowledge and multiple collaborations will move across national borders at warp speed.

Given these circumstances, one must conclude that our educational system is preparing our graduates for a world that ceased to exist some time ago. In addition, enrollment in science and engineering in the United States continues to decrease and the attrition rates are correspondingly high. In a landmark study by Seymour and Hewitt (2000), it was found that students that received degrees in the sciences were similar in abilities to those that had switched majors. Major reasons given for dropping out of the sciences were the poor quality of the teaching, the sheer boredom of the courses, and a perception that they had little relevance to any career that would be of interest to these students.

Under pressure from industry and government, universities and their faculties have begun to face the need for change in their science, technology, engineering, and math (STEM) curriculums. This has been problematic in that many of these efforts focus on the restructuring or creation of a course by an individual professor. This does not necessarily lead to a revision in a course of study nor in the development of a process for sustainable change. Not to mention that dissemination and adoption by other institutions happens rarely and in a random manner.

A NEW EDUCATIONAL FRAMEWORK: CONCEPTS AND POINTS TO CONSIDER

The fundamental change therefore is to realign the courses with the world of work which university graduates will enter. A novel educational framework for STEM would enable students to learn how to acquire and use an ever-expanding body of knowledge where change is occurring at breakneck speed. At the same time, it must expose students to the dynamics of a diverse population, work in teams, and globalization. In a nutshell, it should enable them to work and live in a changed world.

We will present a holistic curriculum that is composed of two bridging concepts. The first one is the "Virtual Workplace" that provides students with a spectrum of thought processes and skills that prepares them for a variety of scientific and science related careers. The second concept is "Journey without Maps." It addresses the challenges associated with the increasing diversity of our student body and faculty. For many minority students, finding an educational pathway through a puzzling and complex university or college system is indeed a journey without maps. For nonminority students, using their education and skills in a culturally heterogeneous and constantly changing global economy is also a journey without maps.

The change of an existing STEM curriculum into a "Virtual Workplace" requires us to consider three educational elements: content/process, skills, work environment.

> **Content/process**: The focus should not be so much the learning of a certain body of information. It should rather be the learning of information in relationship to its use for the solution of major scientific problems. The students should be encouraged to seek information from multiple sources including texts, primary papers, laboratory manuals, the Internet, dialogue with specialists. The information should be reviewed critically and be

interdisciplinary in nature. And the student should understand that the information will continue to grow and change, and that he/she will continue to learn throughout his/her working life.

Skills: The classroom environment should provide an opportunity for students to learn and practice certain fundamental skills, e.g., critical thinking, teamwork, peer review, experimental manipulations, computer use, scientific writing, and oral presentations.

Work environment: The tasks assigned in class should mimic those in the workplace, e.g., a paper describing a project should approximate the format of a scientific publication or a grant proposal, class work should be organized around student teams, a project might yield more than one technical solution or that solution might be an imperfect one though an improvement on previous knowledge.

"Journey without Maps" addresses the issue of how to practice science in a global environment which often involves the interface between science, economics, and culture. At the same time, the students will face the challenge of working in teams that will be diverse in terms of gender, race/ethnicity, class, and educational background. How does one design a course so that it can effectively deal with:

Globalization: The scientific topics can be presented in the context of different social, economic and cultural environments. For example, immunological assays represent an excellent solution for the detection of HIV in blood samples. This procedure is less satisfactory in developing countries due to reasons of cost, availability of medical personnel, and cultural resistance to drawing of blood. This leads to the development of alternative technologies for working with urine or saliva samples.

Diversity in the workplace: The educational process should expose students to the experience of working together with students of diverse backgrounds. This should result in a rational process for arriving at a consensus and maximizing the contributions of every member of a team. The final outcome should be representative of a team effort. Role playing can be invaluable in exploring the value systems of a different group.

Assessment and evaluation: This is an integral component of course and curriculum change, both as a measure of the effectiveness of the innovations and also as a means of maintaining quality control over time. This can be done in a manner that is built into the course by tracking performance with the increasing difficulty in the tasks and by exit surveys of the students. The far more difficult evaluation involves the impact of the new courses on performance in senior level courses and in studies/work following graduation.

The concepts and course features described here are designed to give the student the experience of how a scientist works and thinks in the context of various career tracks. The transformation of the classroom requires a serious consideration of the points described above. Such an initiative runs counter to the existing culture in most universities. First, changes occur mostly at the level of individual courses not of courses of study. Rather than rethinking all the features of a course, it usually addresses one or two elements (e.g., introduction of problem sets, new experiments in the lab). Second, courses that are student centered change the role of the teacher from being master of the classroom to that of a facilitator or arbiter. Third, the teacher becomes the architect and builder of the new course with the resulting investment of time and effort. Fourth, active learning and teamwork increase the difficulty in assessing student performance and put the teacher in the position of having to deal with personality conflicts in dysfunctional teams. Most faculty members are ill prepared to deal with such problems, and in some cases, they may have chosen science as a way of avoiding such conflicts.

THE PRACTICALITIES OF IMPLEMENTING CHANGE

The concepts of a "Virtual Workplace" and a "Journey without Maps" may provide answers for our traditional STEM educational approach. They might even be exciting and intellectually challenging but at the end of the day, we have to get real. There are real constraints. Senior administrators may be supportive of STEM reform but they will warn that it must be done in a resource-neutral manner. The budgets will remain the same. The demands for teaching time by faculty will also not change. The objective, however, is to establish a process that will lead to comprehensive and sustained change across a series of courses even in the face of such constraints. And as in the case of quality research, this process should be faculty initiated.

Given these fundamental concepts and the set of constraints, the question is how can they be implemented at a research university. This section describes a case study that involves microbiology courses at the University of Maryland, College Park, with the participation of roughly 10 faculty members over a period of 15 years. The overall scheme allows for the development of different courses for various student populations.

- Honors seminars: These are interdisciplinary, cross-cultural courses with a maximum enrollment of 20 of the university's best students. These seminars represent a test bed for the development of new educational approaches and teaching materials. If change does not work with very intelligent and highly motivated students, it is unlikely to work with the average student population.

• Lower level, large enrollment science courses: In many respects, these courses are built around adaptations of what has been learned in the honors seminars and reach out to the mainstream of the student body.

• Lower level, general education courses: These courses represent adaptations for nonscience students and are directed at improving science literacy and providing an understanding of the culture of science.

• Upper level science courses: These are the specialized courses for majors and represent a ramping up of the tasks embodied in the concepts of a "Virtual Workplace" and a "Journey without Maps."

This array of courses (and students) enables the creation of a sturdy platform that uses developments in one course to be adapted and applied to other ones. While the objective is to come up with a number of constructs that are applicable to all of these courses, we have found that large-scale introductory lecture/laboratory courses represent a major challenge of their own. For example, the honors seminars are highly effective in their use of student-developed case studies, the use of mixed student teams, and role playing; in a seminar on Traditional Chinese Medicine as a Complementary Approach to Modern Western Medicine, teams may examine the process of scientific and clinical validation as applied to acupuncture for pain management or the use of specific herbal formulations for chronic conditions such as arthritis or dermatitis. However, those course characteristics are only applicable to small classes (i.e., 20 students in the seminars). Major elements such as teamwork and case studies must be adapted for large introductory courses. The following issues, while applicable to all courses, had special difficulties as applied to the introductory courses.

1. How can a course be designed to be interdisciplinary, provide a window to how scientists work, and give a sense of different career opportunities? The basic mechanism is a course module that is presented over a period of several weeks. The module integrates a series of lectures, a case study, mini-quizzes, and a series of laboratory experiments. The case study provides a narrative and a major research question, and the student team needs to find information from multiple sources in order to resolve it. In a semester, the three modules can provide an insight into three different career directions: bacteriology, genetic engineering/biotechnology, and pathogenesis/medicine.

2. How do you construct the course so that it integrates learning of basic concepts, research, and laboratory methods? Each module synchronizes a set of activities (lectures, readings, mini-quizzes, and laboratory experiments). The case study defines the scientific problem which is then broken down into smaller bite size elements. Information from the various activities needs to be accessed and integrated to resolve the case study. This involves a series of mini-quizzes leading up to a paper at the end of the module and a test. The students learn that different types of information are needed and that only some of it is derived from the textbook. The solutions generated by each team may vary.

3. How can students learn the basic skills that are needed for scientific careers? It is generally accepted that knowledge of various laboratory manipulations and familiarity with scientific equipment are an important component of STEM education. There are other skill sets that are equally important and should be built into the courses such as experimental design, team work, computer skills, communications (oral and written), and critical acquisition of information.

4. How can issues of diversity and globalization be addressed? A diverse workplace presents both opportunities and risks which cannot be ignored. The use of teams that are mixed by gender, race/ethnicity, field of study, and grade point average provides a venue for experiencing diversity. Two important elements in our construct have been the inherent difficulty of tasks (requiring maximum effort by every member of the team), evaluation of the task as a team effort, and, finally, peer review in the final grading. The idea is that the more effective the team, the better the outcome of a project whether in the lab or the preparation of a paper. One major aspect of globalization is in the way that modules and case studies are constructed to give a broader perspective, e.g., immunomodulators derived from ethnobotany as an alternative to chemically synthesized drugs as a solution to infectious diseases.

The case study provides support for a pedagogical platform that implements the concepts presented earlier and operates within the constraints of our administrative system. The modification of a set of courses requires components that are, however, not entirely within the domain of faculty members and yet are essential for the success of the enterprise. One of these is evaluation and assessment. Our efforts have focused on building part of the evaluation process into each course. Each successive task in a course is ramped up in difficulty so that proficiency at each stage is necessary to do well in the next one. Class performance in a novel course is compared with that of the traditional version and student surveys are conducted at the end of the semester. Positive results provide some measure of the success of the reforms. Our teaching team feels reasonably satisfied that it has developed a functional model for a large enrollment lecture/lab science course. An early evaluation shows a much higher degree of satisfaction with the new course as compared with its traditional counterpart. Student performance is as good or somewhat better.

We do believe that far more valuable indicators would be performance in successive upper level courses and, ultimately, in graduate/professional school or the workplace. Such projects are clearly beyond the capacity of faculty members or even individual departments.

As pointed out earlier, innovations in the STEM curriculum are expected to be resource neutral, both as regards budget and faculty time. In our case, the solution has been in the use of course design, teaching teams, and technology. Course design incorporates team projects, self-assessment, and peer review which reduces the amount of faculty time involved in grading. In the large introductory course, we have used teaching teams composed of faculty who are responsible for lectures, teaching materials, exams, and overall grading; graduate TAs who deal with the labs and grading of quizzes and exams, and most importantly, undergraduate TAs who act as facilitators and resource persons (most often in relation to questions arising from the modules and case studies). So while overall staffing has increased, this has not had a major impact on budget. Undergraduate TAs are not paid but receive credits for their time. While faculty time has not increased in a major manner, it probably results in an increase of 2-4 hours/week. The course changes cannot be accommodated in the time allotted to lectures and labs. The use of WebCT allows for a 24/7 access to information and ongoing discussion and access to the members of the teaching team. Student difficulties with concepts or scientific details can be monitored, leading to real-time adjustments in lectures and lab sessions. Finally, we have made extensive use of university services: computer expertise (from Office of Instructional Technology), access to information (Library Services), and faculty development and assessment (Center for Teaching Excellence). The use of undergraduate TAs and university services increases the effective manpower without affecting the course budget.

INDIVIDUAL INITIATIVES, SYSTEMIC CHANGE

Changes in the STEM curriculum are typically the result of efforts by individual professors and groups of faculty. The biggest challenge still remains and that is systemic change in a campus and dissemination across institutions. As described above, major elements of curriculum change need to be part of the administrative framework in order to maintain momentum and have sustainability. Assessment and evaluation require resources and expertise that are usually not available to an individual professor or department. Furthermore, the procedures should be common to a college if not to an entire university (possibly through a campus wide Center of Teaching Excellence). The creation of a course of study involves several linked courses. Both the knowledge base and skill sets would be ramped up over a period of three years. Such an effort would require the coordination of content, case studies/problem-based learning, and strengthening of work skills across courses. We are just beginning to do this with a group of faculty that teaches the principal courses in our microbiology curriculum.

While curriculum changes are supposed to be financially neutral, the cost and effort for reshaping or creating a new course does require additional funding. Most often that comes from external grant funding. These grants are usually for two years while the process of establishing a new course and integrating it into the curriculum is more in the range of three to five years. And as teaching assignments are rotated, there is no provision for faculty development as new instructors are assigned to a course. The funding cycles are not well synchronized with curriculum change.

Even as the curriculum of study for a given major or department undergoes major restructuring, there is seldom a process of harmonizing this across the various departments or colleges that are responsible for STEM teaching. And beyond this is the process of dissemination across different institutions. One significant national effort has been a summer institute organized by the National Research Council and the University of Wisconsin–Madison and supported by the Howard Hughes Medical Institute. The purpose of this five-day institute is to bring together faculty teams from various universities to learn new pedagogical approaches to undergraduate STEM teaching. Similar workshops are regularly organized by organizations such as Project Kaleidoscope and the American Society for Microbiology. These activities serve to stimulate grassroots initiatives by faculty. There is little evidence that they lead to systemic change.

A highly educated and skilled workforce lies at the heart of an advanced post-industrial society. Therefore, effective and efficient teaching should have pride of place in our universities and colleges. This paper has argued that we have an increasing understanding of the concepts and tools that can be used for the creation of effective courses and that this can be done in different types of institutions. Such efforts require ingenuity, energy, and time that are comparable to those that go into quality research. Unfortunately, the recognition and rewards are not comparable. Creative and sustainable change cannot be based solely on the initiative and effort of individual faculty but must be sustained by radical change in the administrative structure and reward system of our universities and colleges.

There are a number of possibilities as regards systemic change. These include:

- The creation of a new institute designed to carry out basic research, graduate training, and undergraduate teaching. The University of Basel (Switzerland) created the Biozentrum which was central to the creation of a new Biology II undergraduate curriculum.
- The establishment of a model undergraduate curriculum that includes textbooks and laboratory experiments which is then disseminated to other universities in a national system. The University of Wuhan (China) is doing this in microbiology under the auspices of the Ministry of Education.
- The creation of a new technology university incorporating both new faculty and curriculums. Hong Kong built the Hong Kong University of Science and Technology along the lines of a U.S. research university.

These efforts share certain common characteristics: there is a political will that taps into human and financial resources at a regional and, most often, at a national level. The creation of large new institutes or universities also allows for changes in promotion systems and financial rewards. While there may be analogous initiatives in the United States, our nation differs from other advanced industrial countries in that it does not have a centralized system of education. To put it another way, our system is positioned for innovative approaches to student learning but lacks a framework for sustained and systemic change. This country lacks a lead institution or partnership that can mobilize ideas and resources at a national level. Neither the National Science Foundation nor the Department of Education has undergraduate STEM education as a principal component of its portfolio. The absence of a national system does not preclude the creation of a systemic organization for STEM reform that includes its major stakeholders such as educational institutions, government, and industry (both high-tech employers and those that play an important role in education such as publishing, media, and software). Individual initiatives are all important, but the time has come for systemic development and implementation.

Acknowledgments: The work described in this paper was done in collaboration with a group of energetic and committed faculty and staff members at the University of Maryland, College Park, among which are Ann Smith, Richard Stewart, Patricia Shields, Jennifer Hayes-Klosteridis, Paulette Robinson, Bonnie Chojnacki, Maynard Mack, Jr. (former director of the University Honors Program), and Spencer Benson and James Greenberg (director and former director of the Center for Teaching Excellence, respectively).

Appendix E

Questions to Guide the Review of Undergraduate Food and Agriculture Programs

This checklist of questions is intended to be used by any individual or group conducting a review of any program, curriculum, department, college, or institution. It is designed to assist a variety of organizations in developing specific review criteria, accreditation standards, etc. that incorporate the elements of undergraduate education discussed in this report.

The committee also hopes that this list of questions can guide the assessment of outcomes that follow in response to the report. For example, the elements in this checklist could serve as the basis for follow-up conversations and meetings about undergraduate education in agriculture.

The committee does not suggest what might be the "correct" answers to these questions as the most appropriate responses will depend upon the unique strengths, opportunities, and missions of particular institutions, colleges, and departments.

CURRICULUM AND STUDENT EXPERIENCES

How is the curriculum developed? What is the role of faculty and students within the department? Within the college? Outside of the college? How are external stakeholders engaged?

How do courses in the major build a deep foundation of factual knowledge, based on clear conceptual frameworks?

How does the curriculum incorporate courses and/or experiences focusing on teamwork and working in diverse communities, working across disciplines, communication, critical thinking and analysis, ethical decision-making, and leadership and management?

How are food and agriculture integrated with general education and courses outside of the college of agriculture? How many courses are cross-listed with departments outside of agriculture, especially at the introductory level?

How are real-world examples, case studies, and opportunities for community engagement and service learning integrated into the curriculum?

How do the curriculum and other learning experience reflect contemporary issues and emerging trends in food and agriculture? How are newly arising issues integrated into the curriculum?

In what ways do required courses help students acquire habits of disciplined learning, intellectual curiosity, independence of mind and critical thinking, follow trains of reasoning, detect fallacies in arguments, and discern unstated assumptions?

What levels of international experience associated with global food and agriculture does the curriculum provide/require of students? Which learning abroad opportunities are available and how many students participate? How are international perspectives included in the curriculum?

What opportunities are available for students to participate in internships, cooperative education experiences, service learning, or mentorships? Are any such experiences required?

In what ways are undergraduate students engaged in outreach and extension activities?

What opportunities are there for students to be involved in learning communities or other extracurricular activities that support learning? Are any such experiences required?

What are the opportunities for students to engage in undergraduate research? What percentage of students do so?

INSTITUTIONAL COMMITMENT TO TEACHING AND LEARNING

What faculty development resources and opportunities are available at your institution? What training is made available to new faculty and others offering instruction? What institutional resources are available for developing or refining new courses?

How are faculty encouraged to participate in educationally focused seminars and workshops within your institution? Outside your institution?

How often do seminar and colloquium speakers at your institution discuss issues of teaching and learning?

What is the common method of instruction used in courses? Where on Bloom's Taxonomy of Learning (Bloom et al. 1956) is most instructional effect directed? How are active and cooperative learning integrated into courses?

What forms of instructional technology are used in courses? What institutional resources are available to assist faculty in the use of technology?

How are graduate students and postdoctoral researchers engaged in undergraduate education reform efforts at your institution?

What is the role of teaching evaluations? What elements are included? How are the evaluations used by administrators and others?

What resources are available for bring instructional technology into the classroom?

How many faculty members conduct research on teaching and learning within the discipline?

How are teaching and learning incorporated into considerations for hiring, promotion, and tenure?

OUTREACH AND ORGANIZATIONAL STRUCTURE

How are business, industry, government, nongovernmental organizations, farmers, and community and consumer groups engaged in the development of the curriculum?

What is the composition of any advisory boards with responsibility for food and agricultural education?

How often do faculty members collaborate with researchers and practitioners from outside of academe?

How often do faculty members spend sabbaticals outside of academe? How often do professionals from the food and agriculture industry and other sectors teach courses at your institution?

What types of connections and interactions does your institution have with other academic institutions in the region? Are there joint programs, shared resources, or other types of partnerships in food and agriculture?

What types of articulation agreements does your institution have with community colleges and other institutions within the region?

What types of programs directed at K–12 students does your institution offer?

What types of connections and interactions does your institution have with K–12 students and teachers? With area youth-focused programs such as 4-H, National FFA, and scouting?

REFERENCE

Bloom, Benjamin S., David R. Krathwohl, and Bertram B. Masia. 1956. *Taxonomy of Educational Objectives: The Classification of Educational Goals*. New York: David McKay Company.

Appendix F

Committee and Staff Biographies

James L. Oblinger (*Chair*) is Chancellor of North Carolina State University, North Carolina's flagship university for science, engineering, and technology. NC State has grown under his leadership, operating on an annual budget of $1.04 billion and a $544 million endowment with nearly 8,000 full-time employees and 32,800 students. Since arriving at NC State in 1986, Chancellor Oblinger has served as associate dean and director of academic programs in the College of Agriculture and Life Sciences, dean of the College of Agriculture and Life Sciences, and provost and executive vice chancellor for Academic Affairs.

As Chancellor of NC State, he has worked to make higher education affordable for low-income students and their families through the creation of NC State's Pack Promise; raised $1.3 billion in additional funding for the creation of new facilities and campus improvements through the Achieve! capital campaign; created international partnerships with a number of educational and exchange opportunities for faculty, students, and business executives; spearheaded innovations in teaching and using new technology to improve learning; supported multidisciplinary programs that meet evolving needs of the 21st century, such as the Golden LEAF Biomanufacturing Training and Education Center, which provides trained workers for the state's growing biotechnology industry; and advocated for students and their needs, co-editing an e-book, *Educating the Net Generation*. Under his leadership, NC State's Centennial Campus was recognized as the 2007 Top Research Science Park by the Association of University Research Parks and continues to be a model for innovative partnerships between government, business, industry, and higher education.

Chancellor Oblinger also serves in a number of organizations, including the J. William Fulbright Foreign Scholarship Board, the National Association of State Universities and Land-Grant Colleges Board of Directors, and the American Council on Education Commission on the Advancement of Racial and Ethnic Equity. He also has received several awards for teaching and educational excellence, both as a faculty member and administrator. Dr. Oblinger is a member of the American Association for the Advancement of Science, the American Council for Science and Health, the Council for Agricultural Science and Technology, and the Institute of Food Technologists. He also is a member of Alpha Zeta, Gamma Sigma Delta, Phi Beta Kappa, Phi Epsilon Phi, Phi Kappa Phi, Phi Tau Sigma, and Sigma Xi.

Chancellor Oblinger received his B.A. from DePauw University in bacteriology and his M.S. and Ph.D. in food technology from Iowa State University. Prior to his arrival at NC State, Dr. Oblinger was associate dean and director of Resident Instruction in the College of Agriculture at the University of Missouri–Columbia, as well as professor of Food Science and Human Nutrition at the University of Florida. He is an expert in the microbiology of red meats and poultry, decontamination techniques, and food-borne pathogens.

John M. Bonner is Executive Vice President of the Council for Agricultural Science and Technology (CAST). Before coming to CAST in 2005, he spent 15 years at Land O'Lakes Purina Feed (LOL) LLC, first as beef production manager and then as beef production and marketing manager. He also served as LOL training and marketing manager and eastern sales manager. His efforts at LOL included the introduction of new technical materials and sales support videos, which increased sales in all regions; development and marketing of the "A Steak in the Future" program, which increased LOL beef sales 91%; and the creation and implementation of increased training for sales staff, with an increase in staff members from 20 to 55. Prior to LOL, he worked in the animal health industry in research, training, and marketing. In 2001, Dr. Bonner was named a Fellow of the American Society of Animal Science. He is a member of several professional societies including the National Cattlemen's Beef Association and the Iowa Cattlemen's Association. Dr. Bonner received his Ph.D. from Iowa State University with a nutrition physiology major and economics and physiology minors. He has extensive experience in supervising and encouraging staff and coworkers and is proficient in both development and implementation of successful, profitable agricultural programs.

Peter J. Bruns is the Vice President for Grants and Special Programs at the Howard Hughes Medical Institute (HHMI). Prior to this appointment,

Dr. Bruns was a professor of molecular biology and genetics at Cornell University. Dr. Bruns has earned a national reputation for his efforts to improve science education for students at all levels and oversees a national portfolio in undergraduate science education at HHMI. At Cornell, Bruns established a number of innovative science-education programs, including the Cornell Institute for Biology Teachers, which brings New York State high school teachers together each summer for lectures, field trips, hands-on laboratories, and computer training to improve their teaching of molecular biology. He was a member of the NRC committee on design, construction, and renovation of laboratory facilities. Dr. Bruns received his Ph.D. from the University of Illinois and his B.S. from Syracuse University.

Vernon B. Cardwell is the Morse-Alumni Distinguished Teaching Professor in the department of agronomy and plant genetics at the University of Minnesota. Dr. Cardwell teaches primarily undergraduate courses in grain and seed technology, crop growth and development, crop management, and biology of food, land and the environment. He also provides leadership in educational programs and is very active in K–16 food, fiber, environment, and natural resources literacy efforts. In addition to his teaching efforts, Dr. Cardwell also serves as advisor for students majoring in Applied Plant Sciences, Ag-Industries and Marketing, and all minors in Agronomy. He is a member of graduate programs in Applied Plant Sciences and Conservation Biology. He currently serves on the AAAS Education Committee, and was recently elected to the National Board of Directors for Food, Land, and People. He recently published an article on "Content Standards for Agriculture or Agriculture Content Imbedded within Core Standards" in *Agricultural Education Magazine* and "Literacy: What Level for Food, Land, Natural Resources, and Environment?" in the *Journal of Natural Resources and Life Sciences Education*. Dr. Cardwell received his Ph.D. from Iowa State University.

Karen Gayton Comeau is president emerita of Haskell Indian Nations University. Prior to this position, she directed Haskell's teacher-training program and chaired its teacher education department. Dr. Comeau has also taught at Huron College in South Dakota, the University of Utah, and Arizona State University. During her 11-year faculty appointment at Arizona State University, she was the Director of the Center for Indian Education and editor of the *Journal of American Indian Education*. Dr. Comeau has devoted her career to improving educational opportunities for American Indian/Alaska Native students. Her research at the University of Utah has been instrumental in the recognition of learning styles as an important element in the professional development of pre-service and in-service teachers in schools attended by American Indian and Alaska Native children. Dr. Comeau is a member of the Standing Rock Sioux Tribe who was born and raised on the Standing Rock reservation in North Dakota. Dr. Comeau received her Ed.D. in educational administration from the University of North Dakota. She holds an M.S. in elementary school administration and a B.S. in elementary education from Northern State University in Aberdeen, South Dakota.

Kyle Jane Coulter is a former Deputy Administrator for Science and Education Resources Development in the Cooperative State Research, Education, and Extension Service (CSREES) of the U.S. Department of Agriculture (USDA). In administering higher education programs at CSREES, Dr. Coulter's collaborative efforts with key leaders at colleges of agriculture was instrumental in advancing faculty competencies, strengthening curricula and experiential learning, and attracting academically talented and multiculturally diverse students into food and agricultural sciences undergraduate and graduate degree programs. Nationally, her responsibilities included designing and launching innovative programs that resulted in increased support for state and local agricultural education programs. She broke new ground in reaching out to engage the full system of U.S. colleges and universities and has been especially successful in partnering with minority-serving institutions. Early on, she challenged Colleges of Agriculture to undertake systemic reform in order to do a better job of producing society-ready graduates by addressing food and natural resource systems in the context of human health and welfare, environmental integrity, global competitiveness, and economic security.

In 1993, Dr. Coulter was awarded an Honorary Doctor of Laws degree by the University of Arizona for her achievements in fostering change in higher education in the food and agricultural sciences at the national level. In 2001, she received a Presidential Rank Award (Distinguished Executive) from President Bush in 2001 for being a driving force in uniting USDA and the university system in a campaign to recapture excellence in higher education in the food and agricultural sciences. In 2002, the Future Farmers of America selected Dr. Coulter to receive a special VIP Citation for making significant contributions to agricultural education.

Susan J. Crockett is Vice President and Senior Technology Officer, Health and Nutrition at General Mills, where she directs the Bell Institute of Health and Nutrition. General Mills is the sixth largest food company in the world and has headquarters in Minneapolis, Minnesota. Since 1999, Crockett has been responsible for health and nutrition strategy and programs for General Mills' businesses, health, and nutrition regulatory affairs and issues management, external representation, nutrition science including dietary intake research, and health professional communication.

With support of a Bush Foundation Leadership Fellowship, she completed a Ph.D. in epidemiology from the University of Minnesota in 1987. She has B.S. and M.S. degrees in nutrition and dietetics, is a registered dietitian, and is a Fellow of the American Dietetic Association.

Crockett was Dean of the College for Human Development at Syracuse University from 1990 to 1999 and prior to that was a Department Chair, faculty member, and Extension specialist in nutrition at North Dakota State University. She has published research about nutrition education in schools, effectiveness of nutrition interventions in rural medical clinics and communities. She writes about the influence of environments on the eating behavior of children. In 1987, Crockett received an award from the Secretary of Health and Human Services for Innovation in Health Promotion and Disease Prevention for her proposal, "Parent Health Education: Maximizing Impact." Her research has been funded by the Retirement Research Foundation and the National Institutes of Health (NHLBI and NCI) and she has consulted for the Centers for Disease Control and Prevention, Division of Nutrition and Division of Adolescent and School Health.

Crockett is president-elect of the Board of Directors of the International Food Information Council, a member of the Food Forum that advises the Food and Drug Administration, is active in the International Life Science Institute, helps lead the Minneapolis United Way's Early Learning Initiative, and is a Trustee of the United Theological Seminary in New Brighton, Minnesota.

Theodore M. Crosbie is the vice president of global plant breeding at Monsanto Company. Dr. Crosbie is a seed scientist by training and holds a Ph.D. from Iowa State University. He was recently appointed to a four-year volunteer position as Chief Technology Officer by Iowa Governor Vilsack to coordinate the execution of a three-part economic development "road map" to enhance Iowa's biotechnology, advanced manufacturing, and information technology sectors. Crosbie also serves on the executive committee of the Biosciences Alliance of Iowa, a nonprofit organization formed one year ago to implement the recommendations of Battelle's biosciences report, which was released in March 2004. In 2002, he was named a Distinguished Science Fellow in recognition of his service and management at Monsanto.

Levon T. Esters is an Assistant Professor of Agricultural Education and Studies at Iowa State University. He has several years of experience coordinating pre-college career development programs focused on the agricultural sciences for urban high school age youth. Dr. Esters is a Wakonse Teaching Fellow and a certified Global Career Development Facilitator. His research interest focuses on the career development of students enrolled in secondary and postsecondary programs of agriculture. In particular, he specializes in the application of social cognitive career theory to diverse youth in urban life science educational contexts. Dr. Esters serves on the editorial board of the *Career and Technical Education Research Journal* as well as the Editing Managing Board of the *Journal of Agricultural Education*. He is a member of several professional societies including the American Association of Agricultural Education, the National Career Development Association, and the Society for Vocational Psychology. One of Dr. Esters' most significant accomplishments includes the development of a survey instrument measuring agriscience education self-efficacy which has been used in several studies with students across a variety of cultural contexts (i.e., urban, rural, Korean, New Zealand, and U.S.). Dr. Esters received a Ph.D. in agricultural and extension education from Pennsylvania State University, an M.S. in agricultural education from North Carolina A&T State University, and a B.S. in agricultural business from Florida A&M University.

A. Charles Fischer is the former President and Chief Executive Officer of Dow AgroSciences LLC. He also recently retired from the chairmanship of the Dow AgroSciences Members Committee, which is the executive board overseeing policy and investment for Dow AgroSciences. Mr. Fischer has extensive international experience, especially in Europe, the Middle East, Africa, and Brazil. He was a resident of both Brazil and France during his career at Dow AgroSciences. He served in a leadership role for the Central Indiana Life Sciences Initiative, and also as a board member of the Biotechnology Industry Organization. Fischer was the first person to serve the agriculture industry as president of both CropLife International and CropLife America. He is also past chairman of the National FFA Foundation. Fischer was named 2002 Agribusiness Leader of the Year by the National Agri-Marketing Association, and has been honored by the Mayor of Indianapolis for his leadership in the area of disability awareness. Mr. Fischer grew up on a dairy farm near Cuero, Texas, and earned a bachelor's degree in animal science from Texas A&M University.

Janet A. Guyden is the Associate Vice President of Research and the Dean of Graduate Studies at Grambling State University. Prior to this appointment, Dr. Guyden was the interim dean for the College of Education and a Professor of Educational Leadership at Florida A&M University where she served as coordinator of the educational leadership doctoral program, interim department chair, and the associate chair of the department. Her research interests include the impact of organizations on individual functioning with specific interest in Historically Black Colleges, assessment and program evaluation, and teacher education reform. Dr. Guyden received her Ph.D. in educational leadership from Georgia State University, her M.Ed. in counselor education from Worchester State College, and her B.A. in English from Howard University.

Michael W. Hamm is the C.S. Mott Professor of Sustainable Agriculture at Michigan State University. Dr. Hamm received an Innovation and Leadership Award from the Mid-Atlantic Food and Farm Coalition for his longstanding commitment to the agriculture community and significant contributions to food and farming in the Mid-Atlantic region. Dr. Hamm's research and outreach is focused around community-based food systems and community food security. He also works to identify opportunities for farmers and consumers to link in socially/economically constructive ways. Within community food security his efforts are focused around insuring that all community residents obtain a culturally acceptable, nutritionally adequate diet through a sustainable food system that maximizes community self-reliance and social justice. The C.S. Mott group he heads is focused on three main areas of activity: small- and medium-scale family farm viability; equal access by all members of a community to a healthy diet; and dispersing animals in the countryside. He was a past dean of Academic and Student Programs at Cook College at Rutgers University. Dr. Hamm received his Ph.D. in nutrition at the University of Minnesota and his B.A. in biology at Northwestern University.

Michael V. Martin is chancellor of Louisiana State University. He previously served as president of New Mexico State University (NMSU) from 2004 to 2008. Before coming to NMSU, he served for six years as vice president for agriculture and natural resources at the University of Florida, leading the university's Institute of Food and Agricultural Sciences with more than 3,000 employees statewide. He was elevated to senior vice president of the University of Florida shortly before being selected as NMSU's president. Previously, he was vice president for agricultural policy and the dean of the College of Agricultural, Food, and Environmental Sciences at the University of Minnesota. He began his academic career at Oregon State University as a faculty member in the Department of Agricultural and Resource Economics. Dr. Martin completed a bachelor's degree in business and economics and a master's degree in economics at Mankato State College (Minnesota State University) in Minnesota. He received his Ph.D. in applied economics from the University of Minnesota in 1977. He has been active in professional and community service organizations, including the Farm Foundation's Bennett Agricultural Round Table, the National Agricultural Biotechnology Council, and the Florida Agricultural Resource Mobilization Foundation. He is a member of the American Economic Association, the American Agricultural Economics Association, the International Association of Agricultural Economics, the International Agricultural Trade Research Consortium, the Sigma Xi Scientific Research Society, and the Economic History Association. His areas of specialization are marketing, prices, international trade, public policy, transportation, and business logistics. He continues to be active as a scholar and has written numerous book chapters and articles for academic journals, trade publications, and the popular press.

Susan Singer is Laurence McKinley Gould Professor of the Natural Sciences at Carleton College, where she has been since 1986. From 2000 to 2003 she directed the Perlman Center for Learning and Teaching, then took a research leave supported by a Mellon new directions fellowship. She chaired the Biology Department from 1995 to 1998 and was a National Science Foundation program officer for developmental mechanisms from 1999 to 2001. In her research, she investigates the evolution, genetics, and development of flowering in legumes; many of her undergraduate students participate in this research. She is actively engaged in efforts to improve undergraduate science education and received the Excellence in Teaching award from the American Society of Plant Biology in 2004. She helped to develop and teaches in Carleton's Triad Program, a first-term experience that brings students together to explore a thematic question across disciplinary boundaries. She is a member of the Project Kaleidoscope (PKAL) Leadership Initiative national steering committee and has organized PKAL summer institutes and workshops. At the National Research Council, she was a member of the Committee on Undergraduate Science Education and the Steering Committee on Criteria and Benchmarks for Increased Learning from Undergraduate STEM Instruction and chaired the Committee on High School Science Laboratories: Role and Vision; currently she serves on the

Board on Science Education and is a science consultant to the NRC Science Learning Kindergarten to Eighth Grade study. She has B.S., M.S., and Ph.D. degrees, all from Rensselaer Polytechnic Institute.

Larry Vanderhoef is the Chancellor of the University of California, Davis. He earned his B.S. and M.S. degrees in 1964 and 1965 from the University of Wisconsin, Milwaukee, and a Ph.D. in plant biochemistry at Purdue University in 1969. After one postdoctoral year at the University of Wisconsin, Madison, he was appointed assistant professor of biology at the University of Illinois. He became professor and head of his department in 1977. In 1980, he became provost at the University of Maryland, College Park. Four years later he was hired as the executive vice chancellor of UC Davis and one-person governing board of the UC Davis Medical Center campus in Sacramento. He also served as acting vice chancellor for academic affairs and acting vice chancellor for research. In 1991, after permanently assuming responsibility for Academic Affairs, Vanderhoef was named executive vice chancellor and provost. On April 6, 1994, the UC Board of Regents named Vanderhoef the fifth chancellor of UC Davis.

Chancellor Vanderhoef's research interests lie in the general area of plant growth and development, and in the evolution of the land-grant universities. He has taught classes at levels from freshman to advanced graduate study. Chancellor Vanderhoef has served on various national commissions addressing graduate and international education, the role of a modern land-grant university, and accrediting issues. Chancellor Vanderhoef has been awarded two honorary doctoral degrees, by Purdue University in May 2000, and by Inje University, Korea, in April 2002.

Patricia Verduin is the vice president of global research and development for Colgate Palmolive Company. She was most recently senior vice president and chief scientific officer at at the Grocery Manufactures/Food Products Association. Before that, she was senior vice president and director of product quality and development of ConAgra Foods, where she provided leadership for all research, development, quality, and food safety activities across the organization. Prior to this role, Verduin was a senior member of ConAgra Food Grocery Products' technical team in Irvine, California. Dr. Verduin has held a number of technical positions at Nabisco, International Home Foods, and Lipton. She holds several patents from her research. In addition, she had leadership responsibility for Nabisco's plant operation in Fairlawn, New Jersey. Pat received her B.S. degree from the University of

Delaware in 1980; her M.B.A. in Finance from Farleigh Dickinson University in 1984; and her Ph.D. in Food Science from Rutgers University in 1991.

Dr. Verduin is a member of the Board of Directors for the National Food Processor's Association and sits on the Scientific Affairs Committee for the Grocery Manufacturers of America. She also serves on the Board of Directors of the Alliance for Consumer Education.

STAFF

Adam P. Fagen is a Senior Program Officer with the Board on Life Sciences of the National Research Council. He came to the National Academies from Harvard University, where he most recently served as Preceptor on Molecular and Cellular Biology. He earned his Ph.D. in molecular biology and education from Harvard, working with physicist Eric Mazur on issues related to undergraduate science courses; his research focused on mechanisms for assessing and enhancing introductory science courses in biology and physics to encourage student learning and conceptual understanding, including studies of active learning, classroom demonstrations, and student understanding of genetics vocabulary. Fagen also received an A.M. in molecular and cellular biology from Harvard, based on laboratory research in molecular evolutionary genetics, and a B.A. from Swarthmore College with a double-major in biology and mathematics. In addition to genetics and molecular biology, he is interested in improving undergraduate and graduate science education and other scientific workforce and policy issues. He served as co-director of the 2000 National Doctoral Program Survey, an online assessment of doctoral programs organized by the National Association of Graduate-Professional Students, supported by the Alfred P. Sloan Foundation, and completed by over 32,000 students.

Since his arrival at the National Academies in 2003, Fagen has served as study director for *Bridges to Independence: Fostering the Independence of New Investigators in Biomedical Research* (2005), study co-director for *Treating Infectious Diseases in a Microbial World: Report of Two Workshops on Novel Antimicrobial Therapeutics* (2006), study co-director for the *2007 and 2008 Amendments to the National Academies' Guidelines for Human Embryonic Stem Cell Research* (2007, 2008), study director and co-editor of *Understanding Interventions that Encourage Minorities to Pursue Research Careers: Summary of a Workshop* (2007), and study co-director for *Inspired by Biology: From Molecules to Materials to Machines* (2008). He is currently study director or responsible staff officer for several ongoing projects including the National Academies Summer Institute on Undergraduate Education in Biology, the National Academies Human Embryonic Stem Cell Research Advisory Committee, Research at the Intersection of the Physical and Life Sciences, and Laboratory Security and Personnel Reliability Assurance Systems for Laboratories Conducting Research on Biological Select Agents and Toxins.

Karen L. Imhof has been an Administrative Assistant with the National Academies' Board on Agriculture and Natural Resources (BANR) since October 2003. She previously worked with BANR from 1998 to 2001 as a Project Assistant. For the interim years she was a Senior Project Assistant on the National Academies' Board on Earth Sciences and Resources. Before coming to the Academies, she worked as a staff and administrative assistant in diverse organizations, including the Lawyers' Committee for Civil Rights Under Law, the National Wildlife Federation, and the Three Mile Island nuclear facility and records storage facility. Karen's personal interests include reading, hiking in the woods, jewelry making, and the pursuit of humor.

Robin Schoen is director of the National Academies' Board on Agriculture and Natural Resources (BANR) of the National Academies, a position she assumed in November 2004. Prior to joining BANR, she was Senior Program Officer for the Academies' Board on Life Sciences (BLS), where she directed several studies, including *Discovery of Antivirals Against Smallpox*; *Stem Cells and the Promise of Regenerative Medicine*; *The National Plant Genome Initiative: Objectives for 2003-2005*; *Sharing Publication-Related Data and Materials: Responsibilities of Authorship in the Life Sciences*; and a BANR study on *Predicting Invasions of Nonindigenous Plants and Plant Pests*. She also organized multiple years of proposal and progress reviews for the State of Ohio to assist its efforts to build a biotechnology industry within the state. Before joining BLS in 1999, she worked in various capacities over a 10-year period in the Academies' Office of International Affairs, the National Research Council Executive Office, and the former Commission on Life Sciences. Her work during that time focused on involving U.S. scientists in efforts to strengthen biology internationally, and in addressing policy issues that affect progress in microbiology, neuroscience, biophysics, cancer research, physiology, and biodiversity. This included workshops and reports on gaining access to research resources, intellectual property rights, and developing the infrastructure for science. She also directed a program to bring high-quality laboratory courses in the biomedical sciences to young investigators in Mexico and South America, funded by the Howard Hughes Medical Institute. A native Washingtonian, Robin received a B.S. in biology and chemistry from Frostburg State College, Maryland, and an M.A. in Science and Technology Policy from George Washington University.

Peggy Tsai is a Program Officer with the National Academies' Board on Agriculture and Natural Resources, which she joined in November 2004. She has worked on various studies ranging from agricultural biotechnology to animal health to international agriculture, and served most recently as the study director for *Agriculture, Forestry, and Fishing Research at NIOSH* (2008). She began her work with the National Academies as a Christine Mirzayan Science and Technology Policy Fellow. Prior to this, she interned with the U.S. House of Representatives, House Science Committee; U.S. Department of State, Bureau of Oceans, Environment, and Scientific Affairs; and U.S. Department of Commerce, Technology Administration. Peggy received an M.A. in science, technology, and public policy from George Washington University, and a B.S. in microbiology and molecular genetics with a double major in political science from the University of California, Los Angeles.